MINORITIES IN
SHARK SCIENCE

T0093138

Minorities in Shark Sciences showcases the work done by Black, Indigenous, and People of Color around the world in the fields of shark science and conservation. Edited by three minority researchers, it provides positive role models for the next generation. Highlighting new and important research done in the fields of biology, ecology, and evolution, the book places emphasis on scientists with diverse backgrounds and expertise from around the world.

Despite the use of the term 'Minorities', most of the world's population do not identify as white nor male, and in fact all "minorities" together comprise the global majority of humans. For those in these historically underserved and underrepresented demographics, it is meaningful to be highlighted and be given credit for their contributions. This book showcases to the world the many Black, Indigenous, People of Color, and LGTBQ+ scientists leading marine conservation, both in terms of scientific research and science communication.

It has been shown in the literature that diversity in scientists creates diversity in thought, which leads to innovation. Strong minority voices are exactly what is needed to bring greater attention to the conservation of sharks, and this book illustrates innovative science by people who were historically excluded from STEM. It highlights the unique perspectives these scientists bring to their field that allow them to interact with stakeholders, particularly in the areas of conservation and outreach.

As we continue to amplify these often-forgotten voices through research, outreach and engagement, we hope to stimulate innovation and transformative change in the field of shark conservation and marine science.

MINORITIES IN SHARK SCIENCES

DIVERSE VOICES IN SHARK RESEARCH

Edited by
Jasmin Graham
Camila Cáceres
Deborah Santos de Azevedo Menna

CRC Press
Taylor & Francis Group
Boca Raton London New York

CRC Press is an imprint of the
Taylor & Francis Group, an **informa** business

First edition published 2023
by CRC Press
6000 Broken Sound Parkway NW, Suite 300, Boca Raton, FL 33487-2742

and by CRC Press
4 Park Square, Milton Park, Abingdon, Oxon, OX14 4RN

CRC Press is an imprint of Taylor & Francis Group, LLC

© 2023 Jasmin Graham, Camila Cáceres, Deborah Santos de Azevedo Menna

Library of Congress Cataloging-in-Publication Data
Names: Graham, Jasmin, editor.
Title: Minorities in shark sciences : diverse voices in shark research / Jasmin Graham, Camila Caceres, Deborah Santos de Azevedo.
Description: First edition. | Boca Raton : CRC Press, 2023. | Includes bibliographical references and index. | Summary: "This book showcases the work done by Black, Indigenous and People of Color around the world in the fields of shark science and conservation. Edited by one Black and two Latin American female shark researchers, it strives to provide positive role models for the next generation. Strong minority voices are exactly what is needed to bring greater attention to the conservation of sharks, and this book illustrates how great things have been, are and will continue to be accomplished by people who were historically excluded from marine science. It highlights the unique perspectives these scientists bring to their field that allow them to interact with stakeholders"-- Provided by publisher.
Identifiers: LCCN 2022032057 (print) | LCCN 2022032058 (ebook) | ISBN 9781032196961 (hardback) | ISBN 9781032196947 (paperback) | ISBN 9781003260370 (ebook)
Subjects: LCSH: Sharks--Research. | Minority scientists--Biography. | Minorities in science.
Classification: LCC QL638.9 .M545 2023 (print) | LCC QL638.9 (ebook) | DDC 597.3072--dc23/eng/20220729
LC record available at https://lccn.loc.gov/2022032057
LC ebook record available at https://lccn.loc.gov/2022032058

ISBN: 978-1-032-19696-1 (hbk)
ISBN: 978-1-032-19694-7 (pbk)
ISBN: 978-1-003-26037-0 (ebk)

DOI: 10.1201/9781003260370

Typeset in Minion pro
by Deanta Global Publishing Services, Chennai, India

This book is dedicated to all of the people who have never seen themselves represented in STEM or have ever felt like they are the only people like them in their STEM field. We see you and you are not alone. We also dedicate this book to our ancestors who were victims to colonialism, slavery, racism, and xenophobia. We stand on the shoulders of giants and strive to leave a legacy they would be proud of.

CONTENTS

Preface

WHAT IS MINORITIES IN SHARK SCIENCES?

Minorities in Shark Sciences (MISS) focuses on supporting racial and gender minorities in a focused field. We provide funded opportunities for our members and engage in outreach in broader communities. Our membership is free and open to anyone in high school or older who self-identifies as a racial and gender minority and is interested in a career in shark research and conservation. We were founded by four Black female shark researchers. We strive to be seen and take up space in a discipline which has been largely inaccessible for women like us. We strive to be positive role models for the next generation. We seek to promote diversity and inclusion in shark science and encourage women of color to push through barriers and contribute knowledge in marine science.

Finally, we hope to topple the system that has historically excluded women like us and create an equitable path to shark science. We believe diversity in scientists creates diversity in thought, which leads to innovation.

WHY DID WE WRITE THIS BOOK?

The argument has been made on countless platforms that people involved in certain spaces are chosen for their merit and intelligence, that race and gender have no say in the matter. This begs the following question: Why are science spaces dominated by white men? It is not because minorities lack the merit, intelligence, and drive necessary for these academic spaces; that is simply an excuse to remove such voices from the conversation. This line of thinking is not only harmful to those being excluded but is to the detriment of the spaces from which they are being excluded. For successful discourse on almost any topic, a wide variety of voices must be present for progress to be made—voices that are diverse in race, gender, sexuality, upbringing, socio-economic background, etc. With diversity comes differences in perspective and ways of thinking. That is essential for innovation and scientific productivity.

The term "minorities" can be misleading because, again, minority voices are great in numbers. Some will argue that there aren't enough non-white voices to create change but that is a dangerously false statement. The problem isn't that there are not enough minority voices, but that, historically, there has not been enough representation of these voices, making it even more important to include these voices in every conversation. Minorities are not a negligible part of the population; most of the world's population do not identify as white and are powerful in numbers and thought. For those in these underserved and underrepresented demographics, it is meaningful to see themselves on the stage. To be able to learn from someone you relate to is indescribably influential and will help mold the minds of the younger generations in these communities. That is why we wrote this book. To show the world that not only are Black, Indigenous, and People of Color interested in science and marine conservation, they are making major contributions both in terms of scientific research and science communication. We believe these folks deserve to be celebrated. We also hope to inspire the next generation because science is for everyone.

The work that is desperately needed to change the public's view of sharks is being done right now by many people including many minorities. This strength of minority voices is exactly what is needed to bring greater attention to the conservation of sharks. Now more than ever, those from underrepresented and underserved communities are conducting research, filling knowledge gaps, and teaching their local and global circles about everything they are learning—not just by individuals, but by entire organizations as well, and these organizations are proving the notion that minorities simply aren't interested in science (specifically marine science) wrong. They are doing so by providing a space for representation and recognition of scientists from all different backgrounds.

Editors

Jasmin Graham is a shark scientist and environmental educator who specializes in elasmobranch (shark and ray) ecology and evolution. Her research interests include smalltooth sawfish movement ecology and hammerhead shark phylogeny. She is a member of the American Elasmobranch Society and serves on their Equity and Diversity Committee. She has a passion for science education and making science more accessible to everyone. She is also the Project Coordinator for the Marine Science Laboratory Alliance Center of Excellence (MarSci-LACE) at Mote Marine Laboratory, which is focused on researching and promoting best practices to recruit, support, and retain minority students in marine science. She is also the President and CEO of Minorities in Shark Sciences (MISS) which is dedicated to supporting gender minorities of color in shark sciences. Her work encompasses the areas of science communication, social justice, outreach, education, and conservation. She cares deeply about protecting endangered and vulnerable marine species, particularly elasmobranchs. She works in collaboration with Havenworth Coastal Conservation to study the movements of elasmobranchs in Tampa and Sarasota Bay. Jasmin graduated from the College of Charleston in 2017 with a BS in Marine Biology and a BA in Spanish. She completed her M.Sc. in Biological Science from Florida State University through the National Science Foundation's Graduate Research Fellowship Program.

Camila Cáceres is a marine biologist, shark scientist, and educator. She received a degree in Biology (B.Sc.) from Duke University in 2012 and was first introduced to fisheries research when she was a research assistant at Stanford University's Hopkins Marine Station. She then completed her masters and doctoral studies in the Heithaus Lab at Florida International University. Camila's research focuses on coastal small-scale fishing and coral reef sharks and rays in the Caribbean Sea. She has met with Colombia's Vice-President to discuss marine research, has been featured on Discovery Channel, National Geographic, and Telemundo media among others, and was given the Professional Award at the academic conference Sharks International for her research. As an immigrant, Latina, and LGTBQ, she actively promotes diversity and inclusion in STEM and the outdoors.

Deborah Santos de Azevedo Menna is a marine biologist and shark scientist. Menna has a B.Sc in Biological Sciences with certificates in Geographic Information Systems and Environmental Science at Florida Atlantic University. During her undergraduate years, she participated in several research experiences and trained to become an AAUS scientific diver. She completed a marine science externship with National Geographic Society and The Nature Conservancy. Menna is also a research assistant at American Shark Conservancy (ASC) based in Jupiter, Florida. With ASC she assisted in the study, post-release mortality rate of shore-based angling of Great Hammerhead sharks. She is also a project manager for ASC's "Shark Surveys" project, which is a long-term monitoring program to investigate human-centered and environmental impacts of local shark populations. This is the first project in Florida to utilize any long-term evidence-based methodologies focusing on shark species' diversity and population vulnerability. In December 2021, she was named a National Geographic Young Explorer; this prestigious title is only given to 25 audacious change makers around the globe. Menna is currently working at a respected firm as an Environmental Scientist, conducting surveys on protected species. Her goal is to use her experiences as a multicultural shark researcher and scientist to inspire a new generation of diverse marine biologists from under-resourced and marginalized communities that will champion conservation.

List of contributors

Lauren Ali
The Nurture Nature Campaign
    and Sustainable Innovation
    Initiatives
Port of Spain
Trinidad and Tobago

Miasara Andrew
University of Miami
Miami, Florida USA

Triana Arguedas Alvarez
Minorities in Shark Sciences
Apalachicola, Florida USA

Sara Asadi Gharbaghi
Zwonitz, Germany

Apryl Boyle
El Porto Shark
Los Angeles, California USA

Tatyana Brewer-Tinsley
SeaWorld
San Antonio, Texas USA

Kelly Brown
University of the South Pacific
Suava, Fiji

Camila Cáceres
Minorities in Shark Sciences
Tallahassee, Florida USA

Karla Cirila Garcés-García
Universidad Veracruzana
Xalapa, Mexico

Jaida N. Elcock
Massachusetts Institute of
    Technology
Boston, Massachusetts USA
and
Woods Hole Oceanographic
    Institution
Falmouth, Massachusetts USA

Lara Fola-Matthews
Nigerian Institute for
    Oceanography and Marine
    Research
Lagos, Nigeria

Jasmin Graham
Minorities in Shark Sciences
Sarasota, Florida USA

Ingrid Hyrycena dos Santos
Projecto Tubarão
Santos, Brazil

Aubree Jones
University of Rhode Island
Kingston, Rhode Island USA

Devanshi Kasana
Florida International University
Miami, Florida USA

Salanieta Kitolelei
University of the South Pacific
Suava, Fiji

Gibbs Kuguru
Wageningen University and
    Research
Wageningen, Netherlands

Catherine Macdonald
University of Miami
Miami, Florida USA
and
Field School
Miami, Florida USA

Buddhi Maheshika Pathirana
Blue Resources Trust
Colombo, Sri Lanka

Ilse Martinez Candelas
University of Victoria
Victoria, Canada

Ana P.B. Martins
Florida International University
Miami, Florida USA

Melissa Cristina Márquez
Curtin University
Perth, Australia

Angelina Peña Puch
EPOMEX de la Universidad
Autónoma de Campeche
Campeche, México

Deborah Santos de Azevedo Menna
American Shark Conservancy
Florida, USA

David Shiffman
Life Sciences Center
Arizona State University
Phoenix, Arizona USA
and
Washington, DC, USA

Lauren Eve Simonitis
University of Washington
Friday Harbor Laboratories
Friday Harbor, Washington USA
and
Florida Atlantic University
Boca Raton, Florida, USA

Peyton Thomas
Department of Biology and Marine
    Biology
University of North Carolina
Wilmington, North Carolina, USA

Sabrina Van Eyck
California Academy of Sciences
San Francisco, California USA

Amani Webber-Schultz
New Jersey Institute of Technology
Parsippany, New Jersey USA

Lisa Whitenack
Allegheny College
Meadville, Pennsylvania, USA

Introduction

This book will provide an overview of efforts being made by a diverse group of experts on the research and conservation of sharks and their relatives. Elasmobranchs (sharks and rays) have inhabited our oceans for more than 400 million years, consequently becoming an integral component of marine ecosystems long before human history (Griffin et al., 2008). Sharks are apex predators in the majority of the marine ecosystems they inhabit, and their low abundance is naturally restricted by the ecosystem (Camhi, 1998). Their low numbers, yet wide-reaching effects, give us a hint to the importance of their presence in the oceans.

Elasmobranchs (sharks, skates, and rays) are fish with multiple gill slits with skeletal systems made of cartilage, the same material that the human nose and ear are made of. They are characterized by relatively slow growth, late sexual maturity, and a small number of young per litter. These factors make many shark species vulnerable to overfishing, either captured in directed fisheries or as bycatch in non-directed fisheries. Due to this, approximately one-third of the more than 1000 elasmobranch species are now at risk of extinction, according to a new study re-assessing their IUCN Red List of Threatened Species extinction risk status (Dulvy et al., 2021). Globally, there is a general lack of data reporting on their catch of numbers, particularly species-specific data. For all these reasons, sharks present numerous challenges for their conservation and management.

Marine conservation, also known as marine resources management, is the protection and preservation of ecosystems in oceans and seas. Marine conservation is an interdisciplinary field where individuals engage in science, education, social marketing, economics, resource management, and policy. However, it has developed through a Western knowledge perspective. From early European explorers and collectors to today, management of human relationships with wildlife and wild spaces comes from a Euro-American lens that leaves many voices out of the greater conversation of conservation. But how can those who work in the marine conservation discipline make it more inclusive? Recognizing that colonialism led to imbalances in how ecological research—such as shark research—is collected, produced, and used is an important first step.

By conducting research, assessing their stocks (subpopulations of a particular species), working with fishers, and implementing restrictions on their harvest, many countries worldwide have made progress towards long-term sustainability. However, despite long-standing calls to include minority ethnic groups in research efforts to bring together respectfully and effectively differing viewpoints, traditional knowledge, and Western science, shark conservation is not inclusive and equitable yet. For shark conservation to be truly effective, the field must make the most of such an interdisciplinary and intercultural collaboration by valuing the complex knowledge and understanding of others. It's crucial to understand Indigenous perspectives, including the history and ongoing impacts of colonization.

It is increasingly recognized that different worldviews on human–environment relations are needed to have a "sustainable planet." Many have suggested that to promote inclusivity, changemakers must cut through multiple dimensions of privilege: race, gender, sexuality, nationality, institutional and socioeconomic status, neurodiversity, and even the power of passports. This book offers diverse viewpoints from currently practicing shark scientists conducting groundbreaking research and whose work aims to promote ways of promoting shark conservation and research that is equitable and inclusive.

REFERENCES

Camhi, M. (1998). *Sharks and their relatives: Ecology and conservation* (No. 20). IUCN.

Dulvy, N. K., Pacoureau, N., Rigby, C. L., Pollom, R. A., Jabado, R. W., Ebert, D. A., … Simpfendorfer, C. A. (2021). Overfishing drives over one-third of all sharks and rays toward a global extinction crisis. *Current Biology, 31*(22), 5118–5119. https://doi.org/10.1016/j.cub.2021.11.008.

Griffin, E., Miller, K. L., Freitas, B., & Hirshfield, M. (2008). Predators as prey: why healthy oceans need sharks. Oceana.

1

Public perceptions of sharks

Written by Melissa Cristina Márquez with contributions from Apryl Boyle, Kelly Brown, Jaida N. Elcock, and Salanieta Kitolelei

DOI: 10.1201/9781003260370-1

Melissa Cristina Márquez

In the spirit of reconciliation, I acknowledge the Traditional Custodians of country throughout Australia and their connections to land, sea, and community. I pay my respect to their Elders past and present and extend that respect to all Aboriginal and Torres Strait Islander peoples today.

My name is Melissa Cristina Márquez, and I'm a multi-hyphenate Latina in STEM. Currently a PhD candidate at Curtin University, Perth, Australia, I am interested in what environmental factors influence the composition and distribution of elasmobranchs using a variety of marine technology. I have become a household name thanks to my Scholastic books (the "Wild Survival" series) and TV presenter roles (BBC, Discovery Channel, National Geographic, and OceanX). Host of the *ConCiencia Azul* Spanish podcast, I am passionate about making the scientific industry more diverse and inclusive, including making all of my educational content bilingual. My social media platforms engage a community of over 40,000 through which I encourage meaningful conversations about a variety of topics. Along with monthly articles for Forbes Science, my work has been featured on NPR, Good Morning America, in InStyle Magazine, USA Today, Seeker, Popular Science, and GQ; I am a Forbes "30 Under 30" honoree and was listed as one of InStyle's "BadAss Women for 2021."

Apryl Boyle

I'm a Los Angeles native who was the multiethnic stepchild that shuffled around frequently. The ocean was and is the only place I truly feel at home or find any stability. I double-majored in chemistry and marine science. I received my degree with honors from the University of Tampa, Tampa, USA and began working at Clearwater Marine Aquarium. I was also an Honors Research Fellow and presented work to the American Chemical Society during this time. My research in isomer reactions also resulted in a publication for *Angewandte Chemie.*

After completing summer research in marine chemistry with NOAA when I graduated, I went on to the Medical University of South Carolina and received an MSc in Biomedical Science. My thesis was done in the Department of Biostatistics, Bioinformatics, and Epidemiology and I volunteered at the South Carolina Aquarium during this time. My research on the effects of trichloroethylene on metabolism also resulted in a publication for *Environmental Health Perspectives* and helped to inform EPA policy on Superfund site management. Upon graduation, I came back home to LA and worked as a statistician for a political research firm.

At one point being good at math got me "stuck" in the marketing analytics world, working for companies like Warner Home Video and TBWA\Chiat\Day. Later, I left the security of the big corporations to start my own digital marketing agency, Social Age Media. I secured large clients like Marriott, the Tourism Board of Spain, and Byron Katie International. My lifelong love affair with the ocean continued and I surfed nearly every day. I was involved in volunteer work with Heal the Bay's Aquarium, Surfrider Foundation, and had a stint with the California Science Center's Animal Husbandry team. Finally, realizing that ten-year-old me would be very disappointed that all of the education and experience I had was focused on being a marine researcher but I was doing marketing analytics and strategy instead. After taking an 11-week course in Data Science at General Assembly in the summer of 2015, I was inspired to take the El Porto Shark pet project that was started in 2013 and went full-time into shark and ocean conservation. I've been featured on the Discovery Channel's *Shark Week*, National Geographic, Condé Nast (Vanity Fair and SELF), KTLA Morning News, *LA Times*, OC Register, and other international news outlets. I'm also certified with the International Surfing Association as a surf coach and judging official. Other than marine science and surfing I'm a fan of cooking and hanging out with my dogs. If you really want to be friends you can talk to me about your favorite *Star Wars* movie and/or Star Trek series.

Kelly Brown

My name is Kelly and I am the Curator of the Marine Collection, an aquatic natural history collection at The University of the South Pacific. I am a marine biologist, an avid SCUBA diver, and have been fascinated with sharks since my childhood days. My MSc focused on the life history and diet of juvenile scalloped hammerhead sharks in the Rewa River estuary in Suva, Fiji, and I am currently pursuing a PhD in the population genetic structure and connectivity of holothurians in Fiji. I plan to apply this knowledge to other marine species of fisheries and conservation importance for Fiji and the South Pacific region. The documentation of Indigenous Fijian names of the various marine species that I work with is a recent component of my curator work. I am a Fijian from the province of Cakaudrove, home of the mythological shark god Dakuwaqa and whose people traditionally revere sharks as totems. I have maternal links to the province of Kadavu, home of the mythological octopus-god Rokobakaniceva, who legend says, defeated Dakuwaqa in an epic battle, which afforded the people of Kadavu the protection from shark attacks.

Jaida N. Elcock

My name is Jaida N. Elcock, a desert girl turned marine scientist. The beginning half of my life was spent in the suburbs of Illinois and the second half in the suburbs of Arizona. I spent a lot of my time outside observing the wildlife in the ecosystems around me. Because of this, my family and I knew from a very young age that I would go into a career with animals. As my curiosity grew, I began to ask more complex questions that could only be answered through the scientific method. As a lover of the water, the marine environment was my muse and sharks became my focus as the more I learned how poorly treated and misunderstood they are. But in the desert, there were very limited opportunities for me to interact with the animals I aimed to study. An internship at OdySea Aquarium allowed me to have my first hands on experience with marine animals and it was an opportunity I will be forever grateful for. I went on to obtain my BSc in Biology with University Honors from Northern Arizona University. Following undergraduate, I took part in an NSF-REU experience at Friday Harbor Labs (FHL) with the University of Washington, which provided me my first experience with elasmobranchs in a research setting. At FHL, I had the pleasure of meeting Dr. Stacy Farina and was asked to join her lab at Howard University as a lab technician. This was the best gap year I possibly could have taken, as it gave me the opportunity to continue to build my research skills before becoming a graduate student. Following my year at Howard, I joined the 2020 graduate student cohort at the University of Washington. My first year as a graduate student was riddled with unique challenges, as I began in the middle of a global pandemic. Though I am sad to have missed the full UW experience, I was thrilled to have been given the opportunity to join the MIT-WHOI Joint Program of Biological Oceanography when my advisor switched institutions. I am now a proud student of the Joint Program and I am beyond excited to begin my PhD work.

This work will focus mainly on the biophysical interactions of large pelagic elasmobranchs. I am interested in how they interact with their environments: How are they using oceanographic features such as fronts and mesoscale eddies to their advantage? How are they finding these features? If these features were to disappear in our rapidly changing oceans, how would the sharks need to adjust in order to survive? Many of these questions are largely unknown for the vast majority of elasmobranch species. Answering them would not only help us understand more about sharks and rays, but also about other ecologically and commercially important marine megafauna.

Salanieta Kitolelei

My name is Sala and I am a PhD student at the University of the South Pacific, Fiji. I am a social scientist, farmer, and have always enjoyed listening to stories of sharks from my grandmother. As a child growing up and swimming in the Rewa River, the village elders and mothers always warned me not to swim or make too many splashes while swimming during the shark pupping season, which was signified by the flowering of the blinding trees (*sinugaga*) found along the river. My PhD is focused on using the traditional knowledge of fishers in Fiji and applying it to local solutions for conservation and biodiversity restoration in Fiji. Documenting Indigenous fishing knowledge which elders hold in Fiji is part of my research as I travel across Fiji to collect Indigenous fishing knowledge of the Fijians. I am from the Rewa province, which is home to one of the largest shark nurseries in Fiji. My maternal grandmother hails from Taveuni where Indigenous fishermen use coconut shells to call sharks. My interest in the Indigenous knowledge of fishers in Fiji stems from spending time with my grandparents (both maternal and paternal) and also from field work done with Professor Randy Thaman and an Indigenous master fisher, the late Asakaia Balawa. My research has taught me to pay attention not only to the details being relayed by elders who hold knowledge, but also to their emotions, which cannot be captured through writing. I am privileged to be able to listen and document Indigenous knowledge on sharks, rays and other important species, from elders across Fiji, and honored that the elders entrust their knowledge to me.

KEY TERMS

culling: reduction of a wild animal population by selective slaughter
traditional ecological knowledge (TEK) or local ecological knowledge (LEK) or Indigenous knowledge (IK): the knowledge of local ecosystems obtained by a community through years of keen observation and passed down through generations
endemic: species that exist in only one place in the world
sustainable: able to be maintained at a certain rate or level
trophic: the position an organism occupies in the food web
young-of-the-year: animals born within the past year

INTRODUCTION

In many countries, particularly the Western world, people have a fear which sometimes grows into a hatred of sharks. However, in other countries and cultures of the past and present, sharks are regarded highly and valued as an animal, as a resource, and in some cultures sharks have value in spiritual beliefs, stories, and rituals as well. These positive perceptions of sharks can be overshadowed by sensationalist movies, books, television shows, and even media coverage which continue to depict sharks as man-eating monsters who are a threat to humans. The fear of sharks that has been perpetuated in many countries has led to lackluster support for shark conservation and in extreme circumstances *culling* (purposeful killing of sharks to "protect people"). According to the Florida Museum of Natural History's International Shark Attack File, there were only 57 confirmed shark bites worldwide in 2020, ten of which were fatal. Fifty-seven shark bites with ten fatalities may seem like a lot, but compare that to the 100 million sharks humans kill per year (Worm et al., 2013) and it's clear humans pose a far (ten million times) greater danger to sharks than sharks pose to humans. But how can scientists, conservationists, and policy-makers deal with the threats humans pose to sharks? One step they are taking is changing the perception of sharks. This chapter will discuss how scientists and science communicators in various countries are tackling these issues with a variety of communities.

FIJI: THE INDIGENOUS PERSPECTIVE

When people talk about sharks in an iTaukei (Indigenous Fijian) village in Fiji, the stories revolve around sharks being large scary fish, supernatural guardians, or the sacred fish totem of the paramount chief of Cakaudrove—the Tui Cakau (Morgan). Cakaudrove is one of 14 provinces of Fiji. It occupies the south-eastern third of Vanua Levu (Fiji's second-largest island situated north of Fiji) and has a number of islands close by—including Taveuni (the home of the Tui Cakau),

Rabi, and Kioa. Sharks are culturally significant in Fiji and are revered species living alongside the islanders for millennia that are linked to many iTaukei stories, legends, customs, totems, chants/calling, artifacts, and traditional tattoos and designs. *Indigenous knowledge* of sharks in iTaukei communities across Fiji was developed from fisher interaction with sharks within customary fishing grounds. Indigenous knowledge here is defined as the cumulative knowledge and skills of a community, developed over years of close interaction with their environment and which is passed orally through generations in Fiji. Indigenous knowledge is valuable today because it can fill gaps in knowledge that researchers in fisheries conservation cannot understand and can also provide a basis for a scientific investigation. In this section, the cultural significance of sharks in the iTaukei culture will be discussed. Secondly, it discusses what scientists have found out about shark species in Fiji in relation to their general ecology, and how it complements the knowledge of traditional fishermen and elders as obtained through interviews and observations in Fiji. Thirdly, the section discusses threats to species and opportunities for *sustainable* management from the perspective of Indigenous people. Finally, it addresses how the integration of Indigenous knowledge and science can complement each other, and how this may inform conservation efforts in Fiji and the Pacific.

CULTURAL SIGNIFICANCE OF SHARKS

Historical records written from Fiji in the 1800s include many accounts of sharks, particularly in the shark-rich waters of the Rewa River (MacDonald) and cultural links through totemism. In Fiji, the paramount chief of the province of Cakaudrove and first Tui Cakau, is believed to be born as a twin to a shark god. Fishers from Rukua, Beqa (an island in southern Fiji), heard of the birth of the twins and traveled to the Yavusa (tribe) Benau in Taveuni, which later became the *gonedau* (traditional fishermen) of the Tui Cakau. A close kinship exists between the people of the Cakaudrove province, Rukua village in Beqa, Yanuca Island in the Serua province, the people of the Kadavu Island and of Lakeba Island in the Lau province, all of whom are believed to be safe from shark attacks unless they offend the *vanua* (traditional kinship or relationships in the village). Sharks are considered sacred protectors and guardians of sacred fishing grounds or chiefly households of Cakaudrove province. Anecdotes from witnesses describe sharks saving them from shipwrecks, escorting funeral flotilla of the Tui Cakau (Veitayaki), and patrolling the coastal waters of Somosomo, Taveuni during the mourning period for the paramount chief (Mataitini November 1, 2021, pers. comms to SK). In 1981, an entourage carrying the Tui Cakau's first grandchild from Rewa were escorted by a large shark and school of smaller sharks into the Somosomo Bay (Bukarau April 16, 2021, pers. comms). In Totoya (Lau) a sacred passage called *Daveta Tabu* is believed to be guarded by a shark and if anyone misuses the fishing passage without proper request is liable to being bitten by the shark guardian (*Cinavilakeba*). Shark calling rituals are performed on special occasions by traditional fishing clans or traditional priests (*bete*). The shark calling ritual which involved chanting used to be performed by the *bete* from a clan in Lakeba, Lau, between October and November. The last successful shark call was performed in 1948 by a chiefly lady, who advised her people that they no longer perform the ritual as Christianity prevailed in Fiji, and the elders of the traditional priestly clans no longer performed the traditional ritual. In Taveuni, the *gonedau* led by the Tunidau (head of traditional fisher clan) perform the *qiri qio* ritual on the fourth night after the installation of the Tui Vuna and they use coconut shells from the *niu damu* (Cocos nucifera var) which have been dried for four days to attract sharks ashore (Mataitini and Mataitini 1 November 2021 pers. comms). In Koro, during the traditional turtle calling ritual, a shark usually follows the turtles into the bay where the turtles float up to be viewed. These anecdotes confirm the mana of sharks in Fiji which still exists today in the communities with kinship ties to Cakaudrove, Kadavu, Serua, and Lau.

SHARK TAXONOMY (SCIENTIFIC AND TRADITIONAL TAXONOMY) IN FIJI

According to Seeto and Baldwin's checklist of finfishes identified from Fiji in 2010, there are 61 species of sharks belonging to 15 families (of which 11 species and one family are misidentified or unconfirmed). Since then, scientific research on sharks and their relatives in Fiji has increased, with new information on species occurring in different parts of the country, population information from investigations using genetics, and nursery studies. Shark tourism and community science efforts, such as the Great Fiji Shark Count 2012–2014, have also contributed to species distribution knowledge in the islands.

The most common sharks in Fiji are the requiem sharks (Carcharhiniformes), comprising 16 species (which include the bull, gray reef, blacktip reef, and white-tip reef sharks) and the scalloped hammerhead shark species. In Fiji, juvenile bull and scalloped hammerhead sharks have been studied in detail. From these studies, scientists now know that the peak presence for newborn bull sharks is in December in river systems, and in February, for newborn scalloped hammerhead sharks in estuaries (Glaus et al., 2019; Marie et al., 2017). They also know that juvenile scalloped hammerhead sharks feed on small fish, crustaceans, and small eels (Brown et al., 2016).

Of the total number of shark species identified in Fiji, only 16 species (from five families) are identified by traditional fishers by their local names (Table 1.1) while all other sharks are referred to by their generic name, *qio*. The shark species recognized by local taxonomy are classified mainly by their form, color, habitat, and behavior. The minor classifications used to recognize the shark species include areas they are found, and their perceived danger to humans. According to the iTaukei traditional fishing calendar, sharks give birth between December and January each year and in December, they are more likely to come into contact with people. Elders in communities located along river mouths advise their children and fishers to be careful when out fishing, especially when the blinding tree/blind-your-eye mangrove (*sinugaga—Excoecaria agallocha*) flowers and this tree acts as an environmental indicator for when sharks, most likely bull sharks, swim up river to give birth. According to fishers, when the flower of the *sinugaga* dries and breaks off the tree, the newborn sharks hear this and their eyes open at the sound, get cleansed by the sap from the flower, then they make their way out to sea. Traditional fishers do not target sharks unnecessarily, but only

Table 1.1 Sharks identified by fishers in Fiji with their local names, fisher descriptions and nurseries confirmed in Fiji

Family	Species	Common name	Local name(s)	No. of local names	Fisher description OR translation of name	River nurseries confirmed by shark researchers
CARCHARHINIDAE	Carcharhinus albimarginatus	Silvertip shark	**qio, qio dina, qio seavula**	3	N/A	
	Carcharhinus amblyrhynchos	Gray reef shark	**qio rewasaqa, qio dole, rewasaqa, qio doledole, qio dravu, iko dole, iko uvi, qio saqa, qeo saqa, qio saqa ni cakau, rewasaqwa**	11	Large shark which is grey, with thick torso similar to the trevally (**saqa**) and is very aggressive	
	Carcharhinus falciformis	Silky shark	**qio sisi**	1	N/A	
	Carcharhinus leucas	Bull shark	**qio tovuto, qio dina, qio ni waidranu, iko wawanivata, qio qa, iko yawa, qio ni uciwai,**	7	Large shark with large torso and small eyes. Aggressive and bites when provoked.	Rewa River, Navua River, Sigatoka River, Ba River, Dreketi River
	Carcharhinus longimanus	Oceanic whitetip shark	**qio vulaki**	1	N/A	
	Carcharhinus melanopterus	Blacktip reef shark	**qio mokomoko, qio damu, tutuloa, qio dradra, qio leka, iko tulaniivi, qio tokiloa, qio tulaniivi**	8	Large shark with large torso, gray in colour (check colour) and black tipped dorsal fin	Ba River
	Carcharhinus plumbeus	Sandbar shark	**qio vanuku**	1	N/A	

(Continued)

11

*Table 1.1 (**Continued**) Sharks identified by fishers in Fiji with their local names, fisher descriptions and nurseries confirmed in Fiji*

Family	Species	Common name	Local name(s)	No. of local names	Fisher description OR translation of name	River nurseries confirmed by shark researchers
	Galeocerdo cuvier	Tiger shark	**qio taika, qio daniva, iko vaniva, qio oria**	4	Large shark found in the sea, striped sides similar to the gold spot herring (**daniva**)and bites when it feels threatened.	
	Negaprion acutidens	Lemon shark	**qio damu**	1	N/A	
	Charcharhinidae spp.	Juvenile sharks (gen.)	**qio bulubulu, bulubulu, qeo bulubulu, mata bulubulu (baby)**	4	Juvenile or neonate sharks which have eyes closed and swim up river	
	Prionace glauca	Blue shark	**qio tuiloa**	1	N/A	
	Triaenodon obesus	Whitetip reef shark	**qio, qio dina, leuleu, qio tukivula**	4	NA	
GINGLYMOSTOMATIDAE	*Nebrius ferugineus*	Tawny nurse shark	**qio rukuvatu, qio mocemoce, qio balebale, qio cucu, qio curuqara, qio moce, cave, cuicui, gutumomi, iko mocemoce, mego, qeo cuicui, qeo moce, qio kaboa, qio baleikorokoro**	15	Small shark which is found in sandy areas or hides in rock crevices. Reddish colour with a small snout and does not bit/ attack. **Curuqara**: hides in rocks/caves **Moce**: sleep **Kaboa**:	

(*Continued*)

Table 1.1 (Continued) Sharks identified by fishers in Fiji with their local names, fisher descriptions and nurseries confirmed in Fiji

Family	Species	Common name	Local name(s)	No. of local names	Fisher description OR translation of name	River nurseries confirmed by shark researchers
LAMNIDAE	*Isurus oxyrhinchus*	Mako shark	**qio, qio karawa, karawa, qio mako**	4		
SPHYRNIDAE	*Sphyrna lewini*	Scalloped hammerhead shark	**qio mataitalia, iko ronivi, qio ulutuki, qio ulubatitolu, qio ulutoki, qio colakisi**	6	**Ulutuki:** head shaped like a hammer **Colakisi:** carrying a box	Rewa River, Ba River and Dreketi River
	Sphyrna mokarran	Great hammerhead shark	**qio mataitalia, qio balotu, qio colakisi, qio ulutoki**	4	N/A	Ba River and Dreketi River
STEGOSTOMATIDAE	*Stegostoma fasciatum*	Zebra shark	**qio koba, qio kabo, qio kaboa, ika kabo, ika kabokabo, qeo kaboa, qeo volavola, qio kabokabo, qio kabokaboa, qio koba, qio mocemoce, qio vokai, iko kabo, iko kaboa**	14	Small shark mostly found near the coastline or in sandy areas of the beach, gray in colour with black spots with rough back. Does not bite because of its mouth shape (**gusu qoqo**) and does not taste good when eaten.	
Number of families: **5**	*Number of species:***16**			*Number of local names:***83**		

for special ceremonies. Apart from their belief that sharks are sacred to their culture, they also do not have specific gear or technique to catch sharks unlike the Polynesians that used ropes for shark noosing, along with a rattle lure made from coconut shells. According to an elder fisher who survived a shark attack, sharks are drawn to noise under water like rubbing coconut shells together, the movement of a school of fish, or a swimmer splashing in the water; and the noise also makes them aggressive and prone to attack.

THREATS TO SHARKS AND OPPORTUNITIES FOR USING INDIGENOUS KNOWLEDGE IN MANAGEMENT

Sharks are endangered today because of overfishing, habitat destruction, climate change, and they are also susceptible as bycatch. Shark bycatch is either discarded, eaten, or shared in a community (Glaus et al., 2015). Moreover, because of the value of shark fins and the taste of its meat, sharks are sometimes targeted and either sold to restaurants or consumed. Sharks which live in coastal areas or rivers are highly susceptible to habitat destruction through the removal of mangroves, trawling fishing nets on reefs, land reclamation, and chemical runoff which affect water quality and coral reef health. Climate change raises water temperatures affecting coral health and also increases the frequency and intensity of extreme natural events. These events such as cyclones or flooding alter the river and marine environments which later can impact sharks and their migration routes, feeding habitats and prey availability, and suitable habitats for adults and juveniles. Destructive fishing methods which include coral mining, the use of fish poisons and explosives can also affect shark feeding grounds when their prey are killed or removed affecting the shark's food web. These threats among many others affect sharks in Fiji, and they need to be identified and managed not only using science-based information, but with traditional local knowledge and participation by local fishers. Information on sharks and other important, endangered species should be incorporated into school curriculums, science communication, children's story books, advertisements, and the media to help raise awareness on the threats shark species face and conservation measures.

Using Indigenous knowledge to identify shark species is an important step toward conservation. The correct identification of shark species is vital when dealing with bycatch or shark catch data. Misidentification of sharks sometimes occurs and incorrect data input can provide misleading information on shark fisheries and lead to inaccurate conservation tactics. To counter this misidentification, researchers, fisheries officers, environmental managers, and policy-makers should work with local fishers to add their local and Indigenous knowledge for identifying shark species. Moreover, proper training in monitoring sharks within traditional fishing boundaries is important for the survival of these magnificent creatures to ensure their survival and for our future generations to have a chance at seeing them swimming in our rivers and oceans.

INTEGRATION OF TRADITIONAL SHARK TAXONOMY WITH SHARK SCIENCE—IMPLICATIONS FOR FUTURE RESEARCH IN FIJI

Science and Indigenous knowledge are both valuable on their own; however, when combined, they provide opportunities for learning to many people. A shark research project in Fiji began in 2014 at the University of the South Pacific, and since then, established the gap in baseline knowledge of the shark's basic biological information, spatial and temporal distribution, critical life history, and important habitats for nurseries, foraging, and pupping. Indigenous knowledge of fishers has temporal and spatial information, on which scientists can base their hypotheses about sharks. Scientific and Indigenous knowledge on sharks is discussed but its usually not merged with shark research in Fiji. Because of their daily interaction and understanding of the environment, local fishers can have information which could be used to bridge information gaps and scientific information can inform fishers of nursery and pupping locations which they should avoid to protect sharks from bycatch.

Additionally, restoration of habitats such as mangrove areas and coral reefs are essential activities which should be encouraged among fishers and daily resource users. Future shark researchers should consider bringing local fishers who are shark experts together, and using in-depth interviews, to document as much Indigenous knowledge on sharks as possible. Furthermore, because shark science is in its early stages in Fiji, researchers in the field should not dismiss Indigenous knowledge, but use it to guide their research, particularly ecological and cultural information and knowledge on juvenile species behavior and seasonal indicators. Moreover, although women are forbidden to capture sharks, they should also be included in shark research because women have knowledge about sharks which is different from the men's perspectives.

Just like sharks, Indigenous knowledge of local fishers is in danger of becoming extinct, and both need to be protected. Research collaborations comprising of scientists and local fishers will not only help sharks survive, but also help immortalize our iTaukei fishers' Indigenous knowledge on sharks for future generations to read and to practice.

THE AUSTRALIAN PERSPECTIVE

As a large island nation with a coastline that extends approximately 34,000 kilometers (excluding all small offshore islands), it should come as no surprise that Australia is well known for its beach culture, with a staggering 500 million individuals visiting during 2020–2021 (National Coastal Safety Report, 2021). With one of the longest coastlines in the world, these waters are famous for their turquoise hues, chilly temperatures, and being home to the world's highest diversity of shark species.

There are over 500 species of shark worldwide, with approximately 180 species inhabiting Australian waters and about 70 of those thought to be *endemic* (species that exist in only one place in the world). Sharks are found in all habitats around Australia; however, most are observed on the continental slope or shelf, and a small number are found in freshwater systems (e.g. rivers and estuaries). Most sharks can be legally caught by commercial and recreational fishers, but a handful of species are now listed as "threatened" under the Environmental Protection and

Biodiversity Conservation Act 1999 (EPBC Act). In accordance with the EPBC Act it is an offense to kill, injure, take, trade, keep, or move any member assessed as "threatened" on Australian Government land or in Commonwealth waters without a permit. Under the country's threatened species legislation, any "threatened" species listed can have a recovery plan. Currently, plans exist for nurse (*Carcharias taurus*), whale shark (*Rhincodon typus*), white shark (*Carcharodon carcharias*), and sawfish and river sharks (a multispecies plan). Several shark and ray species found in Australian waters are also protected under the Convention on International Trade in Endangered Species of Wild Fauna and Flora (CITES), which regulates the international trade in wild plants and animals. Additionally, Australia signed up to The Memorandum of Understanding on the Conservation of Migratory Sharks (MoU) in 2011 as six of the seven species covered by this memorandum occur in Australian waters. Fisheries managed by the Australian Government are prohibited to practice shark finning, this being the removal of the fins and discarding of the body of the shark at sea, and similar measures are in place for fisheries managed by the state and territory governments. Fishers are always required to operate consistent with national, state, or territory laws.

SHARKS AND SHARK ATTACKS IN THE AUSTRALIAN MASS MEDIA

Author Peter Benchley has had a lot to do with our fear of sharks—he regrets ever writing the hugely successful book *Jaws*, which resulted in a 1975 blockbuster movie that instilled a deep fear of sharks worldwide that still exists today. Almost 50 years after the release of the film, the perception that sharks are "stalking, killing machines" remains in the public's psyche with new shark movies like *The Meg*, *The Shallows*, and *47 Meters Down* exacerbating those fears.

As most will never encounter a shark in the wild, the media's portrayal of these predators can greatly influence public perception of these species. Sharks are commonly discussed by Australian media outlets and nothing seems to grab the Australian public's attention more than shark and human interactions, making them a popular topic to discuss (Muter et al., 2012). According to the Australian Shark Attack File, in 2020 there were 26 reported shark attacks on humans in Australia (four provoked and 22 unprovoked; Taronga Conservation Society Australia, n.d.). Considering how popular ocean activities are (e.g. beach swimming, underwater diving, surfing), and that a high proportion of this country's population lives near the coast (>80% locals; Chen & McAneney, 2006), these statistics demonstrate how rare human-shark encounters actually are. Yet, when wildlife leads to a human life being lost, and media coverage of the event uses fear-laden language (such as "monster," "attack," and "mindless killer," McCagh, Sneddon, & Blache, 2015), it can generate public anxiety and increase pressure on conservation managers and governments to mitigate risk (Sabatier & Huveneers, 2018).

Despite evidence that many shark species are at risk of extinction (Dulvy et al., 2021; Yan et al., 2021; Pacoureau et al., 2021; Dulvy et al., 2008), scientists observed that most media coverage emphasized the risks sharks pose to people. For example, (Muter et al., 2012) found that shark attacks were one of the most popular article topics in Australian and USA news outlets. Similar results have been found on Australia's mass media Facebook pages (Le Busque et al., 2019; Sabatier & Huveneers, 2018). Facebook is a common way for people to engage with news in Australia and most of the posts from news outlets on this platform discuss "shark/human interactions." Much of the content shared came from television media outlets (1,177 posts), newspapers (568 posts), and radio (273 posts; Le Busque et al., 2019). Le Busque discovered that the majority of shark-related posts came from Australia-wide media outlets (530 posts), followed by media outlets located in Queensland (344), Western Australia (338), New South Wales (334), Victoria (227), South Australia (192), Tasmania (29), and Northern Territory (24).

A common theme from the comments on these Facebook posts became apparent: many believe the ocean to be a dangerous place (n = 2,493), suggesting that people may perceive the risk of shark attacks to be high (Le Busque et al., 2019). However, of the approximately 180 species that call Australian waters home, only three have been responsible for fatal attacks in Australia over the past 20 years—the great white shark (*Carcharodon carcharias*), tiger shark (*Galeocerdo cuvier*), and bull shark (*Carcharhinus leucas*) (Ricci et al., 2016). If a shark species was named in the above Facebook posts, it was most frequently a white shark or tiger shark.

Because human behavior can (and often does) negatively impact shark populations, research that focuses on the social sciences' side of conservation issues is vital in order to create effective management strategies. Understanding the factors behind a community's perception of an animal and creating guidelines is crucial to protecting a species. One such example is the development of suggestions and strategies for how the media should report shark and human interactions. Recently, the Australian media has decided to change their language around "shark attacks." In 2021, two states (Queensland and New South Wales) decided that from now on they will opt to use the words shark "bite," "encounter," or "incident," instead of "attack." Whether other states will follow their example is yet to be seen.

Despite the above, there is ample opportunity for collaborative research between Traditional Owners, scientists, and fisheries' managers in regions where sharks are culturally important. In Australia, there is growing support for the inclusion of IK holders in scientific research and already the field of conservation is seeing what can be achieved from a collaborative approach between Aboriginal and Torres Strait Islander communities and scientists. For example, the Nature Conservancy works together with the Nyikina and Mangala Traditional Owners of the Fitzroy River region in the Kimberley to protect the habitat of the endangered Northern River Shark (Glyphis garricki) and three vulnerable species of freshwater sawfish. Developed by Nyikina Mangala Traditional Owners with support from the Conservancy, the Walalakoo ("a lot of people" in the Nyikina language) Healthy Country Plan implements sustainable environmental practices that are guided by traditional ecological knowledge and kartiya (western) science. This framework integrates biodiversity with the social/cultural values of the Nyikina and Mangala people. While developing ethical and culturally acceptable research may at first be a challenge, shark management could allow Indigenous knowledge holders to participate fully and fairly.

SHARK CONTROL MEASURES AND PUBLIC OPINION

Managing human-wildlife conflict is often challenging and complex. While historically such conflicts were taken care of by killing the "guilty" animal, today many want other alternatives to co-existing with sharks. Shark control measures have been—and continue to be—used in different parts of Australia to quell the country's shark fears. Besides the highly controversial and widely publicized drumlines in Western Australia, where the government sought to catch and kill sharks perceived to be a threat to public safety in 2013–2014, many Australian states today use drumlines and nets. In contrast to studies in Western Australia where support for shark nets was low (~30%; Gibbs & Warren, 2015), there was relatively strong overall support (>60%) among those interviewed on two beaches in New South Wales (NSW) in 2017 for the use of the Sydney shark netting program (Gray & Gray, 2017).

However, despite the above support for netting, NSW beach users were overwhelmingly against (>80%) the general culling of sharks, and also opposed (>70%) the strategy of catching and killing sharks following a shark-bite incident (Gray & Gray, 2017). Since 2015, a majority of respondents have been shown to prefer nonlethal policies (Gibbs & Warren, 2015), a stance that has been recently reconfirmed with respondents having a low preference for lethal mitigation techniques both pre- and post- exposure to the shark media headlines (B. Le Busque, Dorrian, & Litchfield, 2021). Surprisingly, even Australian shark-bite survivors support nonlethal shark mitigation methods (Rosciszewski-Dodgson & Cirella, 2021). In response, new methods and technologies that do not harm marine wildlife are being trialed around Australia. For example, the NSW online community wants "well-reasoned, thought-out [policies] that [are] location-specific and [take] into consideration expert understanding" (Simmons & Mehmet, 2019). The data also showed that the area wants "research energy and resources to be directed to devices [...that are] seen to be in better keeping with community values (safety to all creatures).” New technology, such as drones, as a shark-bite mitigation tool are strongly supported (>85%; Stokes et al., 2020) and are being trialed in NSW and WA. Other beaches currently use a combination of patrol boats, rescue helicopters, and watchtowers to check the waters for sharks.

While the Western Australian shark cull might have been a “turning point in [Australian] public and political attitudes towards shark management” (Fraser-Baxter & Medvecky, 2018), there is still a lot that needs to be done to undo the false narrative surrounding sharks. Despite 37% of the world’s sharks and rays considered “in danger of extinction” as of 2021, the portrayal of sharks continues to be negative, often shown to be more of a threat to Australians (e.g. “shark infested waters”) than the other way around. One step to help sharks would be the government prioritizing the use of nonlethal, scientifically, and conservationally focused management tools. As media analyses have shown, any management strategies should feature consultation with local communities and deliver policies that reflect the values and interests of all Australian people.

THE UNITED STATES PERSPECTIVE

PORTRAYAL OF SHARKS IN THE MEDIA

Sharks seem to be one of the least liked animals in America (Kellert, 1984). A healthy amount of caution should be held with any wild animal, but fear of sharks in the United States seems to be aggressive and contagious. This may be due to a lack of knowledge (for example, reproduction, behavior, movement/migration, habitat use, etc.) and misinformation (Afonso et al., 2020). Shark-related stories shared in and by mass media (e.g. newspapers, television news, news outlets on mobile devices, etc.) are often of negative encounters with sharks, labeling them as vicious “attacks” when that is hardly the case. For example, an article was released in 2021 entitled “Watch speargun angler come face to face with great white shark off NC’s Outer Banks”. With thrashing prey in the water to pique the sharks' interest, it is less than surprising that a spearfisher would encounter a shark. The animal is not chasing the person, it is searching for injured fish. *The New York Times* published an article about the "Fear on Cape Cod as Sharks Hunt Again". While the article includes helpful information about how the growing seal populations are contributing to the rise in shark sightings, it is important to recognize that a number of people will not take the time to consume the entire article, so it is crucial to choose effective and accurate language when describing sharks in the media. Otherwise, the result conjures up unnecessary fear, especially in coastal communities.

These types of headlines and articles are misguided and/or misinformed and are potentially harmful to both the sharks and humans, as these articles disproportionately discuss negative shark encounters and the risk sharks pose to humans (Muter et al., 2012; Hardiman et al., 2020). These negative stories are the bulk of shark-related media viewed by the average consumer, most likely driving their opinion to be negative and fearful.

There are television programs that claim their aim is to reduce the stigma around sharks and advocate for their protection in a rapidly changing and overexploited ocean. For example, Discovery Channel's *Shark Week* is a week-long, shark-centered event loved by millions, but many involved in marine science agree that over the years, its programming has become less than educational. Dr. Lisa Whitenack and colleagues (2021) reviewed *Shark Week* documentaries from the last 32 years and found that *Shark Week* seems to be doing more harm than good as it stands now. For example, despite the shows having featured a number of shark species, they focus mainly on just a few large and charismatic species. Additionally, although shark bites are not always the focus, undertones of risk to humans and fear are common. Furthermore, the majority of people depicted as authority figures in shark science are white men, some of whom are presented as experts but are non-experts, despite the large number of non-white, non-male shark scientists actively conducting research. Though one would expect programming like *Shark Week* to advocate for the protection of sharks, it has been taken over by sensationalism and it now feels more like televised clickbait with a gross lack of appropriate representation.

THE SURFER PERSPECTIVE

While many marine researchers regularly swim and scuba dive in research, there is a growing body of work being performed by surfing marine researchers and within the surfing community. The Center for Surf Research at San Diego State University is one example of academia recognizing the value of the surfing community. In fact, whether they have collegiate training or not, lifelong surfers can be skilled community scientists.

One of the most pervasive questions for surfers from non-surfers is, "Aren't you afraid of sharks?" In Southern California, the surfing culture has been central to outdoor recreation since the 1960s and a significant number of surfers participate in the sport at least weekly (Surfrider Foundation, 2011). Since they can see a lifetime of coastal change that includes sea level rise, construction, conservation efforts, and other local knowledge the team at El Porto Shark asserts that surfers are uniquely qualified community scientists. Understanding and engaging the surfing community in shark conservation will allow localized ownership of conservation. El Porto Shark's mission is ocean and shark conservation through research, education, and action. Based in Los Angeles, the organization's name comes from a local surf break in Manhattan Beach, California where the founder surfed every morning.

In the spring of 2013, several *young-of-the-year* (born that year) and smaller juvenile great white sharks were spotted frequently in the lineup with surfers. The media at the time was a mix of calm science education and more fear-driven reporting; local morning news shows aired features interviewing researchers explaining that there was no imminent danger. They balanced coverage with local surfers either being non-reactive (as they should) or featuring those claiming that "something needed to be done" to protect humans. It was here, inspired by sociology and public policy studies, that the Shark Public Relations (Shark PR) initiative was born. This project provides accessible marine biology and environmental science education along with an appreciation for and an invitation to physically engage with the coastlines for fun and fitness. This public relations initiative was started in an effort to provide sharks with a "publicist" to help better their reputations. This coupled with El Porto Shark's Chemical Oceanography initiative, which will be covered in a later chapter, aims to create more equity in ocean education and recreation. Changing the narrative from humans needing protection from sharks to sharks needing protection from humans is of global significance and requires dissemination of knowledge about ocean science and safety.

Populations of larger sharks have significantly decreased in the past 50 years and many are close to extinction—for example, there are about 70% fewer larger, predatory sharks since the 1970s when the movie *Jaws* was released (Pacoureau et al., 2021). The way the media talks about shark encounters is important to help conservation. If media coverage continues to call every encounter an "attack," it will continue to create more collective public fear. For example, at times there are encounters where there are no negative human and shark interactions, yet they are still classified as an attack.

During the COVID-19 pandemic, the surfing industry enjoyed record growth and an increase in diversity. Enthusiasts from around the world and varying socioeconomic backgrounds are drawn to the sport to test their balance, patience, and skills by engaging in a regular surf practice. In California there are several newly minted organizations, both public and private, whose mission is to teach people of color how to surf and be comfortable in the ocean. These groups are rooted in the work that has been done by the Wahine Project, Women Who Surf group on Facebook/Instagram, Latinx Surf Club, Brown Girl Surf, Salud y Cariño, El Porto Shark, and other smaller movements through the late 2000s to present. Moreover, women and minority surfers are beginning to surf now more than other demographics in the past decade. Coaching programs, surf schools, and retreats designed to meet the wants and needs of women and minority surfers are growing in number and creating greater accessibility to a new influx of surfing enthusiasts. Many doors can be opened by having these more diverse voices in outdoor recreation and marine science. This said, the aforementioned demographic is still only a small portion of the surfing community and not well studied or even catered to.

The connection between surfers and sharks may not be obvious to everyone and the relationship between surfers and sharks is complex. Longtime surfers often tell tales of at least one encounter with a larger shark at some point in their personal history. Some will say that the water sometimes "feels sharky" or that there is a certain "feeling" that they need to get out of the water before they had planned to. While these encounters can be logically written off to nerves about other conditions (bigger waves, stronger current, bad weather, etc.) there is no doubt that surfers and sharks cross paths on a regular basis.

In a preliminary analysis, cataloged shark sightings submitted by the local coastal community to the Shark Research Committee website from 2006 to 2013 were organized into a database. This website has a simple form where people can record any shark sightings they encounter while they're at the beach on a given day. Surfers overwhelmingly responded with greater frequency (49.7%) than those participating in other shoreside activities such as swimming, angling, or splashing in less than a meter of water. Other shark sighting data sets from local anglers and recreational boating companies were later added for additional analysis.

In 2018, El Porto Shark began a survey of local surfers' attitudes towards sharks and whether they were afraid of sharks and if so, did it influence whether they surf or continue to surf if there is a shark near the lineup. Surveys were created in Google Forms and had 12 required questions that were demographic and asked how often one surfed, where they surfed, and whether they were "afraid of sharks". The data skewed heavily to those who were acquainted with the initial research team. In the fall of 2020, the survey relaunched through a partnership with environmental science students from the University of Southern California. Overall, regardless of the respondent pool, over half of surfers report having seen a shark while surfing. Over 60% of respondents said that they were not afraid of sharks and 15% reported that they weren't sure. The preliminary findings suggest that regardless of demographics, surfers understand there are sharks but aren't worried about the sharks bothering them while surfing (Figure 1.1).

From these findings, El Porto Shark is narrowing language to determine the best "public relations campaign" or Shark PR that activates the surfing

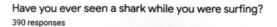

Have you ever seen a shark while you were surfing?
390 responses

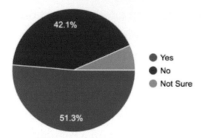

Are you afraid of sharks bothering you while you surf?
390 responses

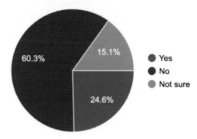

Figure 1.1 *Survey responses from current surfers, mostly in Southern California, asking if they have seen a shark while surfing and whether they are afraid of sharks bothering them while they surf (n = 390; unpublished).*

community as community scientists and shark advocates. They are working with sociologists, psychologists, and policy experts to create a multi-disciplined approach to disseminate knowledge in the area of ocean science to everyone regardless of location, background, education, etc.

The primary work in the Los Angeles area will be to continue to survey and educate long-established and newer surfing groups on ocean and shark conservation. There will also be a focus on partnering with and educating surfers, students, and others to better understand the science that affects their daily decisions. Whenever possible the founding team of El Porto Shark contributes to local science journalism, speaks at schools and events, and has media appearances to explain shark conservation and why everyone should be involved regardless of how far they live from the ocean (Figures 1.2 to 1.4).

Figure 1.2 *Apryl Boyle speaking with media about surfer-shark interactions.*

Figure 1.3 *Apryl Boyle filming on her surfboard near the Huntington Beach Pier, discussing surfer-shark interactions.*

Figure 1.4 *Apryl Boyle surfing at El Porto.*

THE IMPORTANCE OF COMMUNICATING SHARK SCIENCE

Though there are many studies highlighting the ecological, social, cultural, and economic importance of sharks (Bornatowski et al., 2014; Puniwai, 2020), there is a struggle to have this information successfully communicated to the greater public. The best way to combat the misguided fears of the public is through education. Accurate science communication, outreach, and classroom education on sharks—and marine ecosystems as a whole—are essential in creating a more realistic and positive image of these top predators. One study (O'Bryhim & Parson, 2015) shows evidence that people with more knowledge of sharks more readily support the conservation of sharks. Science communicator Scientists and science communicators who share one or more identities with the community they are trying to communicate with can often be more effective in establishing trust with the community. The science shared feels less distant, so much more personal, when one can relate to those sharing the knowledge. Seeing oneself in the experts and authority figures of a field bridges the gap between the community and the field. That is imperative in providing accurate scientific information to all parties and will, therefore, aid in changing the perception of vilified animals, such as sharks, to one that is more positive. Not only that, but the representation of different races/ethnicities, genders, abilities, etc. in science will inspire the next generation of young people from all different backgrounds to become involved in science, which will ultimately change the face of science for the better. Effective science and science communication are not possible without minority voices. It is that simple.

CONCLUSION

There are several factors that were discussed in this chapter that can lead to a fear of sharks including lack of knowledge, lack of exposure to or familiarity with the ocean, and the portrayal of sharks in the media. To address the first factor, it is important that all communities have access to information about the important role sharks play in the ecosystem and how they can impact their daily lives. When people are able to connect with and understand the value of something they are more prone to protect it. People don't want to protect something they feel is dangerous or scary, but they will protect something they see as valuable. Creating a positive public perception of sharks is important, but there also needs to be a positive perception of shark scientists and conservationists so that trust can be built with communities. This is especially important for scientists coming from outside of the community or for scientists who experience many privileges or benefit from the oppressive systems that have historically excluded certain communities from participating in important discussions and policy decisions or from contributing knowledge to the body of information used to make these decisions. There are many communities around the world that value sharks and find them culturally relevant and these communities are important to tap into in order to help drive shark conservation. It is particularly important to engage with Indigenous communities who have generations of knowledge that is, in many aspects, unparalleled.

REFERENCES

Afonso, A. S., Roque, P., Fidelis, L., Veras, L., Conde, A., Maranhão, P., …
Hazin, F. H. V. (2020). Does lack of knowledge lead to misperceptions?
Disentangling the factors modulating public knowledge about and
perceptions toward sharks. *Frontiers in Marine Science, 7.* https://doi.org/10.
3389/fmars.2020.00663.

A Socioeconomic & Recreational Profile of Surfers in the U.S. (2011).
Surfrider Foundation. Retrieved from https://www.surfrider.org/coastal-
blog/entry/a-socioeconomic-and-recreational-profile-of-surfers-in-the-
united-states.

Bornatowski, H., Navia, A. F., Braga, R. R., Abilhoa, V., & Corrêa, M. F. M.
(2014). Ecological importance of sharks and rays in a structural foodweb
analysis in Southern Brazil. *ICES Journal of Marine Science, 71*(7), 1586–
1592. https://doi.org/10.1093/icesjms/fsu025.

Brown, K. T., Seeto, J., Lal, M. M., & Miller, C. E. (2016). Discovery of an
important aggregation area for endangered scalloped hammerhead sharks,
Sphyrna lewini, in the Rewa River estuary, Fiji Islands. *Pacific Conservation
Biology, 22*(3), 242–248. https://doi.org/10.1071/PC14930.

Chen, K., & McAneney, J. (2006). High-resolution estimates of Australia's
coastal population. *Geophysical Research Letters, 33*(16). https://doi.org/10.
1029/2006gl026981.

Dulvy, N. K., Baum, J. K., Clarke, S., Compagno, L. J. V., Cortés, E., Domingo,
A., … Valenti, S. (2008). You can swim but you can't hide: The global status
and conservation of oceanic pelagic sharks and rays. *Aquatic Conservation:
Marine and Freshwater Ecosystems, 18*(5), 459–482. https://doi.org/10.1002/
aqc.975.

Dulvy, N. K., Pacoureau, N., Rigby, C. L., Pollom, R. A., Jabado, R. W., Ebert,
D. A., … Simpfendorfer, C. A. (2021). Overfishing drives over one-third of
all sharks and rays toward a global extinction crisis. *Current Biology, 31*(22),
5118–5119. https://doi.org/10.1016/j.cub.2021.11.008.

Fraser-Baxter, S., & Medvecky, F. (2018). Evaluating the media's reporting
of public and political responses to human-shark interactions in NSW,
Australia. *Marine Policy, 97*, 109–118.

Gibbs, L., & Warren, A. (2015). Transforming shark hazard policy: Learning
from ocean-users and shark encounter in Western Australia. *Marine Policy,
58*, 116–124. https://doi.org/10.1016/j.marpol.2015.04.014

Glaus, K. B. J., Adrian-Kalchhauser, I., Burkhardt-Holm, P., White, W. T., & Brunnschweiler, J. M. (2015). Characteristics of the shark fisheries of Fiji. *Scientific Reports*, *5*(1). https://doi.org/10.1038/srep17556.

Glaus, K. B. J., Brunnschweiler, J. M., Piovano, S., Mescam, G., Genter, F., Fluckiger, P., & Rico, C. (2019). Essential waters: Young bull sharks in Fiji's largest riverine system. *Ecology and Evolution*, *9*(13), 7574–7585. https://doi.org/10.1002/ece3.5304.

Gray, G. M., & Gray, C. A. (2017). Beach-user attitudes to shark bite mitigation strategies on coastal beaches; Sydney, Australia. *Human Dimensions of Wildlife*, *22*(3), 282–290.

Hardiman, N., Burgin, S., & Shao, J. (2020). How sharks and shark–human interactions are reported in major Australian newspapers. *Sustainability*, *12*(7), 2683. https://doi.org/10.3390/su12072683.

Kellert, S. R. (1984). American attitudes toward and knowledge of animals: An update. In M. W. Fox & L. D. Mickley (Eds.), *Advances in animal welfare science, 1984/85* (pp. 177–213). Washington, DC: The Humane Society of the United States.

Le Busque, B., Dorrian, J., & Litchfield, C. (2021). The impact of news media portrayals of sharks on public perception of risk and support for shark conservation. *Marine Policy*, *124*, 104341. https://doi.org/10.1016/j.marpol.2020.104341.

Le Busque, B., Roetman, P., Dorrian, J., & Litchfield, C. (2019). An analysis of Australian news and current affair program coverage of sharks on Facebook. *Conservation Science and Practice*, *1*(11). https://doi.org/10.1111/csp2.111.

Marie, A. D., Miller, C., Cawich, C., Piovano, S., & Rico, C. (2017). Fisheries-independent surveys identify critical habitats for young scalloped hammerhead sharks (Sphyrna lewini) in the Rewa Delta, Fiji. *Scientific Reports*, *7*(1), 17273. https://doi.org/10.1038/s41598-017-17152-0.

McCagh, C., Sneddon, J., & Blache, D. (2015). Killing sharks: The media's role in public and political response to fatal human–shark interactions. *Marine Policy*, *62*, 271–278. https://doi.org/10.1016/j.marpol.2015.09.016.

Muter, B. A., Gore, M. L., Gledhill, K. S., Lamont, C., & Huveneers, C. (2012). Australian and U.S. news media portrayal of sharks and their conservation. *Conservation Biology*, *27*(1), 187–196. https://doi.org/10.1111/j.1523-1739.2012.01952.x.

O'Bryhim, J. R., & Parsons, E. C. M. (2015). Increased knowledge about sharks increases public concern about their conservation. *Marine Policy*, *56*, 43–47. https://doi.org/10.1016/j.marpol.2015.02.007.

Pacoureau, N., Rigby, C. L., Kyne, P. M., Sherley, R. B., Winker, H., Carlson, J. K., ... Dulvy, N. K. (2021). Half A century of global decline in oceanic sharks and rays. *Nature, 589*(7843), 567–571. https://doi.org/10.1038/s41586-020-03173-9.

Price, M. (2021). Watch speargun angler come face to face with great white shark off NC's outer banks. Retrieved from https://www.charlotteobserver.com/news/state/northcarolina/article256813237.html#storylink=cpy.

Puniwai, N. (2020). Pua ka wiliwili, nanahu ka manō: Understanding sharks in Hawaiian culture. *Human Biology, 92*(1), 11–17.

Ricci, J. A., Vargas, C. R., Singhal, D., & Lee, B. T. (2016). Shark attack-related injuries: Epidemiology and implications for plastic surgeons. *Journal of Plastic, Reconstructive and Aesthetic Surgery, 69*(1), 108–114. https://doi.org/10.1016/j.bjps.2015.08.029.

Rosciszewski-Dodgson, M. J., & Cirella, G. T. (2021). Shark bite survivors advocate for non-lethal shark mitigation measures in Australia. *AIMS Environmental Science, 8*(6), 567–579. https://doi.org/10.3934/environsci.2021036.

Sabatier, E., & Huveneers, C. (2018). Changes in media portrayal of human-wildlife conflict during successive fatal shark bites. *Conservation and Society, 16*(3), 338. https://doi.org/10.4103/cs.cs_18_5.

Seeto, J., & Baldwin, W. (2010). A checklist of the fishes of Fiji and a bibliography of Fijian fish. Technical report, 1/2010, University of the South Pacific, p. 107.

Simmons, P., & Mehmet, M. I. (2018). Shark management strategy policy considerations: community preferences, reasoning and speculations. *Marine Policy, 96*, 111–119.

Stokes, D., Apps, K., Butcher, P. A., Weiler, B., Luke, H., & Colefax, A. P. (2020). Beach-user perceptions and attitudes towards drone surveillance as a shark-bite mitigation tool. *Marine Policy, 120*, 104127. https://doi.org/10.1016/j.marpol.2020.104127.

Surf Life Saving Australia National Coastal Safety Survey. (2021). https://issuu.com/surflifesavingaustralia/docs/ncsr_2021.

Taronga Conservation Society Australia. (n.d.). Australian shark attack file. Retrieved from https://taronga.org.au/conservation/conservation-science-research/australian-shark-attack-file.

Whitenack, L. B., Mickley, B. L., Saltzman, J., Kajiura, S. J., McDonald, C. C., & Shiffman, D. S. (2021). *Sharks, lies, and videotape: A content analysis of 32 years of shark week documentaries.*

Worm, B., Davis, B., Kettemer, L., Ward-Paige, C. A., Chapman, D., Heithaus, M. R., … Gruber, S. H. (2013). Global catches, exploitation rates, and rebuilding options for sharks. *Marine Policy, 40*, 194–204. https://doi.org/10.1016/j.marpol.2012.12.034.

Yan, H. F., Kyne, P. M., Jabado, R. W., Leeney, R. H., Davidson, L. N. K., Derrick, D. H., … Dulvy, N. K. (2021). Overfishing and habitat loss drive range contraction of iconic marine fishes to near extinction. *Science Advances, 7*(7). https://doi.org/10.1126/sciadv.abb6026.

2

Elasmobranch ecology and evolution

Written by Deborah Santos de Azevedo Menna
and Jasmin Graham with contributions from
Karla Cirila Garcés-García, and Gibbs Kuguru

DOI: 10.1201/9781003260370-2

Deborah Santos de Azevedo Menna

I was born in São Paulo, Brazil, and at the age of three I moved to South Florida in the United States with my family. I remember when my interest in marine science started to spark, I was in middle school and heard my science teacher talking about the terrible deepwater horizon oil spill. That made me in awe of how powerful, yet vulnerable, the ocean truly is. At that moment I knew I wanted to be a part of making a difference in ocean conservation. I started gaining experience in the marine science field as early as high school. As a freshman I participated in multiple tagging programs with different universities, working side-by-side with scientists to collect data through measurements, tissue samples, and tagging. That is when I had my first encounter shark tagging, and when I saw my first shark in the wild, it was love at first sight! In college I earned a BSc in Biological Sciences with certificates in Geographic Information Systems and Environmental Science at Florida Atlantic University. During my undergraduate years I started training to become an AAUS scientific diver and completed a marine science externship with National Geographic Society and The Nature Conservancy. I also became a research assistant at American Shark Conservancy (ASC, a non-profit founded by Hannah Medd, based in Jupiter, Florida). As project manager for ASC's "Shark Surveys" project I have collected, and am continuously working on collecting, in-water diversity, abundance, and behavioral data from endangered, vulnerable, and near threatened shark species through underwater visual surveys, length measurements, species, and sex determination. In 2021 I was named a National Geographic Young Explorer, and with that opportunity I will be using the grant throughout 2022 to recruit students (ages 17–23, of all genders and identities) from underrepresented communities where they are most in conflict with shark conservation. Through collaborative community science participation with local underrepresented students, they will take part in collecting diversity, abundance, and behavioral data from endangered, vulnerable, and near threatened shark species through underwater visual surveys, length measurements, species, and sex determination. It is important to have diversity in the science field, and to include underrepresented future scientists because diversity in science equals diversity in thought. Participants will engage with seasoned and early career scientists to create a measurable impact on baseline studies for shark biodiversity, which is a community-based and globally relevant marine conservation issue. My dream is to use my experiences as a multicultural shark scientist to inspire a new generation of diverse marine biologists from under-resourced and marginalized communities that will champion marine conservation.

Jasmin Graham

I'm a shark scientist and environmental educator who specializes in elasmobranch ecology and evolution. My research interests include smalltooth sawfish movement ecology and hammerhead shark *phylogeny*. I'm a member of the American Elasmobranch Society and serve on their Equity and Diversity Committee.

I have a passion for science education and making science more accessible to everyone. I'm the Project Coordinator for the Marine Science Laboratory Alliance Center of Excellence (MarSci-LACE) at Mote Marine Laboratory, which is focused on researching and promoting best practices to recruit, support, and retain minority students in marine science. I'm also the President and CEO of Minorities in Shark Sciences (MISS), an organization dedicated to supporting gender minorities of color in shark sciences. I'm excited to help open doors for more underrepresented minority students to join the exciting field of marine science.

My work encompasses the areas of science communication, social justice, outreach, education, and conservation. I care deeply about protecting endangered and vulnerable marine species, particularly elasmobranchs. I work in collaboration with Havenworth Coastal Conservation to study the movements of elasmobranchs in Tampa and Sarasota Bay.

I graduated from the College of Charleston in 2017 with a BSc in Marine Biology and a BA in Spanish. I went on to complete my MSc in Biological Science from Florida State University through the National Science Foundation's Graduate Research Fellowship Program.

Karla Cirila Garcés-García

 I was born in 1987 in the northern part of Mexico, in the state of Coahuila. During my childhood, I moved with my parents to the south of Mexico in the state of Veracruz, where I currently live. I was (and still am) passionate about marine life. I decided to study Biology in Tuxpan, which is a coastal town where fishing is one of the main activities, and sharks and rays are of commercial importance. In 2005 and 2012, I earned my BSc and MSc at Universidad Veracruzana in Tuxpan, respectively. In 2008, the Mexican National Council of Science and Technology (CONACyT) launched a search for students with an outstanding record. I was the first woman from Universidad Veracruzana to be awarded a scholarship for research in the field of commercial fisheries. In 2009 while still an undergraduate, I investigated the growth rate and aging of *Rhizoprionodon terraenovae*, a shark caught in the local artisanal fishery, supervised by Dr Javier Tovar-Ávila. This was remarkable because, at that time, men heavily dominated the field of fisheries research in Mexico, and for a woman, working with fishers was (and is still) challenging in several parts of Mexico due to customs and superstitions. Women close to, or onboard a boat are considered "bad luck" and to bring "misfortune" for fish catches. The conditions of the boats and the pace of work are also extremely difficult. In spite of all this, I demonstrated that a woman could perform scientific fieldwork and develop research communication with fishers under these circumstances.

In 2016, I moved to Australia to undertake a PhD at the University of Melbourne. I was supervised by Dr Robert Day and Dr Terence Walker. Currently, my research is focused on the analysis of ecological vulnerability to the anthropogenic hazards of fishing and climate change for all the sharks, skates, and rays in western Mexico and in the Gulf of Mexico, particularly in Veracruz to contribute to better fisheries. My aim is to evaluate how elasmobranch populations in Veracruz respond to fishing and climate change stressors. I'm currently a professor at Universidad Veracruzana, and mentor marine biology students, as well as advise MSc and PhD students. I collaborate with the National Institute of Fisheries and Aquaculture of Mexico (INAPESCA) on different projects.

The network I have created with different researchers in Mexico has inspired undergraduate and postgraduate students under my supervision to apply for internships and research projects on elasmobranchs in the Gulf of California and the Gulf of Mexico. As in other developing countries, Mexico faces financial issues to support education and research, and without these collaborations many interested students could not have developed their knowledge and applied their skills to research elasmobranchs. I'm currently collaborating with international universities such as Universidad de Antioquia in Colombia, and Universidad San Pablo in Madrid, Spain, to develop a technological application for smartphones to identify sharks and rays in the fieldwork, and to create scientific podcasts to share elasmobranch science.

Gibbs Kuguru

 I was born and raised in Kenya, the country known for the most prolific wildlife in the world. Historically, our nature and wildlife were a key feature of my people's heritage, but in the last century much of it was appropriated and subsequently decimated. Since I was young, I was always fascinated by the wilds of the world and consumed nature documentaries and Zoobooks with a voracious appetite, which has not yet dissipated. This led me to chase opportunities to learn and encounter natural wonders, which eventually led to SHARKS! And with the sharks is most likely where you'll find me. Though, if I'm being honest, I didn't expect this.

While doing my BSc in Arkansas, I got a weekly email from our university bulletin that was advertising a "White Shark Cage Diving Internship" in Mossel Bay, South Africa. Honestly, I believed that such an activity would surely lead to death…after all, Jaws was a man-eating killer, right? However, my alternative option was to try and get into med school for my graduate degree, so I ended up choosing death. To my surprise, I quickly found that sharks not only don't want to kill people, but I'm pretty sure they don't care about humans at all. The significance of being insignificant gave me a cosmic perspective where I was free to be free. Being in their presence became my heart's desire and the focus of my academic interests. Through this, I found a way to marry my love of genetics with my love of sharks and found myself actually excited to do "schoolwork" for the first time in my life. I did my MSc on the genetics of hammerheads and got to work with so many cool folks who worked in research, conservation, tourism, film, communications, law, and more!

My career is based on using genetics to learn about sharks, by revealing the many secrets hidden in their DNA. Currently, my research is focused on understanding the role melanin-producing cells play in the evolutionary survival of these species. It's clear that the shark's skin is a valuable asset that confers not only armor-like protection, but also camouflage to facilitate (or prevent) predation. Sharks that lose the ability to produce melanin pigment through a disease called leucism have a distinct disadvantage that may leave them vulnerable to the many hazards in nature. Many of these disadvantages are worsened by human activities, so it's clear that research without action will not keep sharks, or any other living thing, in existence for much longer.

KEY TERMS

cephalofoil: outward extension of the head of a hammerhead shark

ecology: the study of relationships between organisms and their environment

evolution: the process by which living things develop and diversify over time

effective population size: a value used to calculate the genetic size of a population based on levels of diversity

microsatellite: a series of unique, repeating base pairs that can be used as a genetic fingerprint

phylogeny: the study of how species are related to each other

population: a community of organisms from the same species that live together

SNPs: areas in the DNA where small differences can be found; these small changes are a variation of a single base pair

taxonomist: a scientist that works to classify species based on their shared characteristics

INTRODUCTION

Ecology is the study of relationships between organisms and their environment. There are many different levels at which one can study ecology including organism, population, community, and ecosystem. For example, a clownfish is an organism; all the clownfish on a reef make up a population, and all the different species on the reef make up a community, and where they all live make up an ecosystem. Scientists who study ecology are called ecologists. They ask questions to study the environment, such as: what do organisms need to survive in their environment? What happens when ecosystems are unable to function normally? How do organisms interact with nonliving and living things in their environment? There are many different research areas within ecology such as molecular, population, spatial, community, and behavioral ecology.

Molecular ecologists look at the smallest scale of organisms, how they affect their environment, produce DNA and proteins, how their production can be affected by the environment; and genetic diversity in populations. Organisms have genetic diversity in their populations so they can continue to adapt to certain environmental factors. Population ecologists study the increase and decrease in the number of species. An example is the decrease in the population of great hammerhead sharks (*Sphyrna mokarran*) in the coastal waters of Florida. Spatial ecologists look at specific spatial patterns of different organisms and populations that affect ecological dynamics. It is an important part of ecology because it highlights the study of anthropogenic (human) causes of habitat loss. "Spatial" means relating to or occupying a space, such as the spatial distribution of blacktip sharks (*Carcharhinus limbatus*) during their yearly migration to Florida.

Community ecologists study the biotic and abiotic factors in the ecosystem. Biotic factors are living things such as plants and animals, while abiotic factors are nonliving things such as soil and water. Behavioral ecologists study how an organism reacts to different situations or changes in its surroundings. A behavioral ecologist might study mating patterns, for example. It is important to study ecology because organisms depend on nonliving and living things to survive. By studying how everything coexists, scientists can better understand how to preserve vital resources for clean water, food, and air.

Ecologists are increasingly relying on community science to gather data about the marine ecosystem. Public participation leads to a more informed community that is more supportive of good management. When the public participates, the sharks and our marine ecosystems benefit! Including the general public in research and volunteer work can help them have a better understanding of sharks and assist in their protection.

Evolutionary biology is a branch of science that studies the evolution of species and groups of animals at a variety of time scales. Environmental changes and the interactions within species can cause evolutionary change, and ultimately alter natural selection. Natural selection can change the genes of underlying traits, eventually altering the interaction of organisms with their environment and predators, ushering in a new era of evolutionary transformation. Studying shark genetics can also help scientists understand how they have evolved over time. Genetics is the study of how traits are "handed down" from parent to child, population genetics allows scientists to study the DNA of sharks with more accuracy. It helps understand population demographics, keep track of population health, identify species, and connect populations to one another. Although there are more than 500 species of shark, some sharks look extremely similar or even identical to each other. It can be difficult to differentiate between species of sharks if they have similar physical characteristics. By using DNA barcoding, scientists can differentiate between species. DNA can also be used to determine how closely related species are to each other.

Understanding the differences between different species and the relationships of those species to each other is the first step in assessing the status of an organism. Once scientists can differentiate between species they can start to assess populations of organisms of the same species. To better understand the role certain species play in the ecosystem and the impacts changes in population dynamics might have on the ecosystem, scientists can observe these organisms at the individual level. These individual observations can help scientists learn what an organism eats, where it lives and how it interacts with its environment.

SEEING THE BIG PICTURE: IDENTIFYING SPECIES AND SPECIES RELATIONSHIPS

Sharks are slow growing and slow evolving, but they have existed for a long time, which makes studying their evolution particularly interesting. Sharks, rays, and chimeras, which make up the class Chondrichthyes, are jawed, cartilaginous fishes that first arose in the Paleozoic era (Grogan & Lund, 2004). Most sharks are highly adapted predators and have strong senses including an additional sense that humans do not have, electroreception. Electroreception allows them to sense electrical charges which animals give off to create heartbeats— think of a pacemaker. As the number of other species rose and fell, the shark populations remained steady (Compagno, 1990). Sharks play an important role in marine systems (Compagno, 1990); many are apex predators and help keep fish populations in check, providing balance and stability to the food web (Roff et al., 2016). Sharks have existed on this planet for millennia and have evolved adaptations that allow them to play unique roles in the marine ecosystem. Understanding the evolution of these animals can help scientists better understand their ecological functions.

PHYLOGENY

A considerable amount of information can be learned by studying how closely related various groups are. Phylogeny is the study of how species are related to each other. Like a family tree, a phylogeny shows what species or groups share common ancestors. Phylogenies allow scientists to hypothesize the sequence of changes that likely occurred in a character trait or feature over evolutionary time. In the past, phylogenies were created by looking at the anatomy of organisms, but now scientists are able to use DNA to build phylogenies. In a perfect world, the phylogenies that scientists build from DNA should agree with the phylogenies built from looking at the organism's anatomy, but that isn't always the case. This leads scientists to question…which is correct?

HAMMERHEADS

Enter the hammerhead family, also known as the Sphyrnids. This is one group of animals where the DNA and the anatomy do not agree when it comes to phylogeny. Hammerhead sharks have long intrigued biologists because they have a bizarre head shape, called a *cephalofoil*, which is unique among vertebrates. The origin of this head shape remains a mystery (Kajiura et al., 2003; Lim et al., 2010) and it is not clear what, if any, advantage it gives hammerheads—although scientists have all kinds of theories! Each of the ten species of hammerheads, including the great hammerhead, bonnethead, smooth hammerhead, scalloped hammerhead, and carolina hammerhead, to name a few, has a different shaped cephalofoil (Figure 2.1). If scientists can figure out the phylogeny

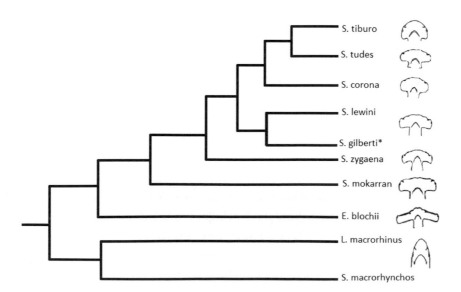

Figure 2.1 *Head shapes of various hammerheads.*

of the hammerheads, they can maybe understand how the cephalofoil is evolving over time which could answer the mystery of, "why the heck are their heads shaped like that!" Not surprisingly, there have already been several attempts to address this question (Compagno, 1990; Naylor, 1992; Quattro et al., 2006; Cavalcanti, 2007; Lim et al., 2010), and yet, we still do not know. Scientists hypothesize that the head has been getting wider over time based on the anatomy (Compagno, 1990), but the DNA tells a different story. In the phylogeny built based on DNA, it seems like the head was an accident, a freak mutation that the hammerheads have been evolving away ever since (Quattro et al., 2006; Naylor et al., 2012). A researcher by the name of Cavalcanti attempted to combine all of the hypothesized phylogenies scientists had built to create one "supertree" (Figures 2.2-2.4). With all of the work that has been done, scientists still do not have a clue of which tree to believe.

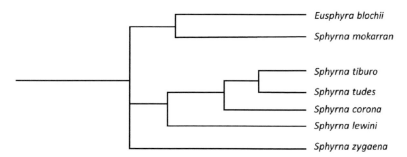

Figure 2.2 *Hypotheses of the phylogeny of the hammerhead family: Based on DNA, proposed by Naylor et al. (2012).*

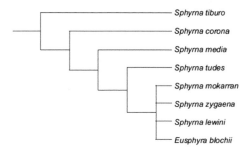

Figure 2.3 *Hypotheses of the phylogeny of the hammerhead family: Based on anatomy, proposed by Compagno (1988).*

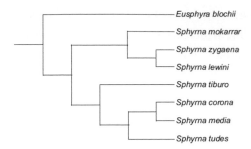

Figure 2.4 *Hypotheses of the phylogeny of the hammerhead family: The supertree combining all hypotheses, determined by Cavalcanti (2007).*

A study was conducted in 2016 by Graham and colleagues (unpublished data) to try another approach to see if the trees would agree. The first step was to rebuild the phylogeny based on the anatomy, but this time they would ignore the head entirely just in case the head was what was leading scientists astray. They also decided to look at the DNA again, but this time use the DNA from the mitochondria. In schools, students are taught that "the mitochondria is the powerhouse of the cell," but many don't know that the mitochondria also has DNA in it. This DNA is particularly unique because it only contains DNA passed down from the mother. Without the mixing of the mother and father's DNA, the DNA of the mitochondria doesn't change nearly as much, so researchers thought it might provide a clearer picture.

HOW DID THEY DO IT?

The team CT scanned an individual from a variety of hammerhead species and created digital three-dimensional models of the individual's anatomy. They then looked for similarities and differences in anatomy across the various hammerhead species. For the DNA part of the study, the team extracted DNA from little pieces of fin taken from sharks during catch and release sampling. The DNA was extracted using a process that consists of adding various chemicals and heating and cooling the sample repeatedly. The DNA could then be analyzed and compared to each other using computer modeling.

STILL NO AGREEMENT

In the end the trees still didn't agree, but they did notice a few things the trees had in common. The first thing they noticed is that in both trees there were two distinct groups of hammerheads: one made up of small sharks with small heads, the other made up of large sharks with relatively large heads. What the researchers discovered is that the trees are really similar aside from the fact that it isn't clear where the tree starts. In other words, the trees disagree on which species is the oldest or most closely related to the first hammerhead. That is what is referred to in phylogenetics as a "rooting problem" and it is quite common. Typically rooting problems occur when scientists try to build phylogenies for large groups of animals that aren't very closely related or evolved so many years apart that the models that scientists use aren't as effective (Stefanovic et al., 2004; Hillis, 1998; Zwickl & Hillis, 2002; Pollock et al., 2002; Rannala et al., 1998). There are a few ways scientists can go about addressing these rooting problems, but for now, the evolution of the hammerhead family remains a mystery. While it may be easy to go along with the latest technology, even DNA analyses have their limits sometimes, especially when dealing with several species that have been around for millennia. But understanding long-term evolution of a group of animals is just one piece of the puzzle. It tells scientists about the big picture of the past and how species are related to each other, but it's also important to think about population dynamics on a species level.

UNDERSTANDING POPULATION DYNAMICS

When the question arises, "How do you keep track of a *population*?", one's go-to method might be to count them using a traditional census method. Theoretically, this is totally possible, but not very easy (Luikart et al., 2010). John Shepherd said it best, "Counting fish is like counting trees, except they are invisible and they keep moving." In practice, this method is cumbersome especially when using a tally to count animals that are difficult to find (invisible) and span the breadth of the ocean. Still, shark scientists needed a way to get reliable information from these populations without having to physically see every single individual. The study of trait inheritance in populations really allowed us to better understand the state of these animals by studying their DNA (Ovenden et al., 2016). These advances in knowledge around DNA birthed a new field called "population genetics."

Genetics is the study of how traits are inherited from parent to offspring, which is an examination of traits across generations. By focusing this examination of the species on the inhabitants of a particular region or community, scientists can study the genetics of the population. The value of population genetics is that it uses a fraction of the individuals required for a traditional census-based study, without losing accuracy. The utility of the population genetics field was further developed by DNA sequencing technologies which gave rise to population genomics. By taking tiny pieces of tissue from a specimen, billions of copies of DNA are readily available to analyze and interestingly enough, it's also much cheaper than a census-based study (Carrier et al., 2019). Since its development, shark scientists have used genetics/genomics to better understand population demographics at the level of species identity, population connectivity, and also population health (Dudgeon et al., 2012).

SPECIES IDENTIFICATION

Taxonomy is a field that is specialized in classifying species by arranging them into categories of similarity. *Taxonomists* developed a set of rules that would classify species based on their shared physical features, but they encountered some problems. As they applied these rules, they found that some species didn't fit into the categories they originally intended, which is what is called the "species problem" (Hey et al., 2003). As you can imagine, this created some confusion, and this was also noted in shark research. Scientists had to innovate new approaches that would help them understand what defines a species, by looking at a variety of different sources of information. This primarily included ecology and biology, but also genetics (molecular taxonomy). This approach is called "integrative taxonomy" and genetics has contributed a lot to this study (Padial et al., 2010). DNA barcoding is a genetic technique used to address the "species problem." DNA barcoding uses a short segment of DNA that has a high degree of similarity within a species, and a low degree of similarity between different species (Ward et al., 2005). This allows the scientist, without specialist knowledge of taxonomy, to easily differentiate two specimens, even without looking at them! In some studies, processed fish products have been taken from markets and restaurants and have been easily identified to their species (Fields et al., 2015). This is a type of DNA forensics that focuses on the illegal practices of wildlife trafficking and is incredibly useful in enforcing conservation of wild animals. While this technique is not perfect, it does show how the inclusion of different sources of data can help clear up the big questions about which animals exist in our oceans and markets.

POPULATION CONNECTIVITY

A population can be defined as a species group or community that live together in an area. Many factors contribute to how connected these populations are, which include oceanic barriers like land masses and currents (Kuguru et al., 2019). The ability for sharks to cross these barriers is something called "dispersal." For example, highly migratory sharks are capable of swimming through different oceanic barriers, which allows them to interbreed with populations in other areas; thus, making them more connected. Given this, their genetic profiles will be more similar to the individuals in their home population than to the other populations (Pardini et al., 2001).

On the other side of that spectrum, there are sharks who never leave their home areas and will reside there for the duration of their lives! (Chin et al., 2012). These populations have less connectivity to other members of their species living in different areas. This phenomenon has been studied using genetic markers like *SNPs* (pronounced "snips") and *microsatellites*, which are areas in the DNA of an animal with a high degree of variation that allow scientists to construct unique genetic profiles between populations (Portnoy & Heist, 2012). Remember, DNA barcoding only goes down to the species level, so these markers have a higher resolution, but it also takes a bit more time to develop and use. Studies have applied these markers to understand the genetic relationships between sharks living in different areas so they can figure out how best to conserve them. An example is that of the great white sharks in South Africa that spend some time with their Australian counterparts (Bonfil et al., 2005). We can see that they are definitely distinct from each other based on their genetic profile, but they also share some genetic characteristics that give them some population level similarity. Based on this, these two populations can be considered as a single conservation unit because their life histories are intertwined to some degree based on how the genes flow between each. However, this is difficult to implement because the politics that govern the conservation practices often can't agree on how to proceed. One government might support a regular culling practice, where sharks are killed to reduce their population numbers (Gibbs & Warren, 2014). Killing (and not conserving) these animals can tear the genetic fabric that makes the population strong and can have a serious impact on the overall health of the population; a type of damage that is not easily undone.

POPULATION HEALTH

The health of a population can be defined as it is evolutionary potential to survive long enough to successfully pass its genetic information to the next generation. As a general rule of thumb, this potential is measured by genetic diversity, where a higher diversity should confer a higher evolutionary potential for survival for the population (Hughes et al., 2008). If all the members of a population are genetically similar to each other, then one can say they are effectively a smaller population. Another way of thinking about this is that the genetic count would be smaller than the census count. This is a measure termed *effective population size* (Charlesworth, 2009). In the presence of natural selection, a low effective population size is susceptible to being defeated by a single environmental stressor that can wipe out the population (think climate change). For example, what if the seas get too warm and the population all has genes for low heat tolerance? This would spell disaster for the delicate ecosystem balance. Inversely, when the effective population size is large, then the possibility increases that a group of individuals in a population has genes for high heat tolerance. This means that these specific individuals of the population will pass their genes to the next generation and help maintain the balance of the ecosystem. Unfortunately, declines in global shark populations and a lack of proper conservation measures (Ward-Paige & Worm, 2017) means that even though population numbers are recovering and may even reach a decent census size, their population health is fragile and can easily succumb to environmental pressures. Ideally, protecting a population and its environment is one way to allow nature to increase the effective population size, but with the chaos stemming from human activities this is not always likely. In some cases, a little bit of intervention is needed to help nature back on its way.

Genetics cannot only be used to identify species relationships and understand population dynamics, it can also be used to understand how species interact with the environment and how they move.

UNDERSTANDING HOW SPECIES INTERACT WITH THEIR ENVIRONMENT

ENVIRONMENTAL DNA

Understanding a day in the life of a larger shark is difficult. White sharks, for example, can travel up to 50 miles in one day. To find out where a large shark spends its time, scientists have long relied on the tag and track method. In this method, the scientist removes a shark from the water, brings the animal onto a boat, then attaches a tracking tag, which may vary in size and functionality depending on the study, to their body. Drones and other aerial and marine visual surveying instruments can also be used to track shark movements. Each of these has limitations and can be constrained by budget.

To identify changes in migratory behaviors, scientists need to be able to detect animals along these migration routes. Simpler, more accessible technologies that can be rapidly deployed with researchers or trained community scientists can collect more data points than researchers alone. By decentralizing and demystifying scientific knowledge, especially in the environmental sciences, communities can own and manage their own natural resources such as nearby coastlines.

Using environmental DNA (eDNA) technology, it is possible to determine whether certain species have recently been in a particular area. People leave DNA in their environment every day. Hair, skin, and other cells are naturally shed, which is one way crime scene technology can tell whether a person has been at a particular location. Sharks are no different. Skin and other cells naturally shed off sharks just like they shed off humans and ocean water can be filtered in order to isolate shark DNA out of the cells that they leave behind. The limitation with this technology is that it can only detect whether an animal has been in the area recently, within an hour or so, due to the nature of the analysis and its currents that move water and everything in it.

A company called Cramer Fish Sciences/GENIDAQ supplies filtration supplies and the sampling method for collecting DNA from ocean water. Samples were collected and shipped overnight to their lab in Sacramento. In order to perfect the methodology, they started by detecting either white or mako shark DNA. The method allows for the detection of any shark species with an available control. In addition, the presence of sharks' favorite prey items are tested and detection of these species is also reported (Figure 2.5).

Community involvement in shark conservation is necessary for successful conservation. This work sheds light on why ocean and shark conservation is important for any healthy ecosystem. This El Porto Shark project (previously discussed in Chapter One) was started in Los Angeles County and engages everyone, including those on the coast as well as inland, to understand what the testing parameters are. First, researchers must understand what the coastal communities actually need, including cultural and regional considerations. The methods developed as part of

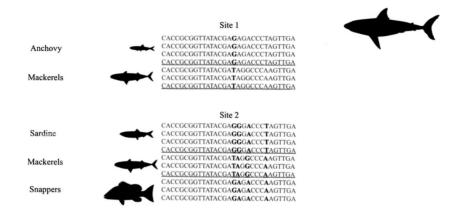

Metabarcoding detected anchovy and mackerel at site 1 and sardine, mackerel and snapper at site 2.

Figure 2.5 *DNA Sequencing for prey species.*

this project require limited space and resources, making it possible to consider rapid deployment of testing supplies to areas where education and conservation work is done.

CLIMATE CHANGE VULNERABILITY STUDIES ON ELASMOBRANCHS

The fact that elasmobranchs have survived the five great extinctions on Earth has allowed people to think that elasmobranchs, as a group, are resilient enough to survive any disaster, including climate change (Joel, 2019). Chondrichthyans are extremely important for marine and coastal ecosystems as some perform as top predators to regulate other individuals in the marine ecosystems. For example, the white shark and the tiger shark regulate seals and marine bird populations, respectively. Additionally, more than 40% of elasmobranchs are of commercial importance and many people, particularly those living in coastal areas, rely on them as a source of protein and income. To understand, however, why chondrichthyans are vulnerable, it is important to consider the life-history traits that they have and to do studies in the field to understand how changes in their habitats affect them. The perception of chondrichthyans as organisms that are sufficiently resilient may need to be adjusted in order to ensure the persistence of their populations during the current sixth extinction crisis (Chivian & Bernstein, 2010; IPCC, 2014) (Figures 2.6 and 2.7).

One way to know how and which chondrichthyans are affected by factors related to climate change is by knowing their vulnerability. Timmerman (1981) gave important insight in climate change topics, and defined vulnerability as the degree to which a system acts adversely when a hazardous event occurs. This definition was used as the basis for climate change analysis and was adopted by others such as the Intergovernmental Panel on Climate Change (IPCC), which is an important working group to study climate change around the world. Years later, the IPCC (2007) defined vulnerability as the degree to which a system is susceptible to and unable to cope with the adverse effects of climate change. That means if something has certain characteristics that do not allow them to

Figure 2.6 (a) Photo of a researcher collecting stingrays onboard a fishing vessel and (b) a full body photo spotted eagle ray.

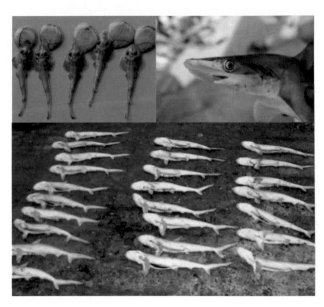

Figure 2.7 *Collage of photos including embryos of a guitarfish (top left), a juvenile hammerhead shark (top right), and several gray sharpnose sharks (bottom).*

cope with adverse factors, then its vulnerability will be high. Of course, vulnerability can change over time based on changes in which hazards occur or if the adverse factors disappear.

A few studies in Mexico, and in other parts of the world such as Australia have shown results of vulnerability analysis on chondrichthyans to climate change stressors in order to know which species are at high vulnerability and need urgent management strategies (Knip et al. 2010 and Martínez-Candelas et al. 2020). Currently, there are more studies of vulnerability analysis from fishing stressors, but it is crucial to assess marine organisms in an integrated way. Thus, in this section the focus will be only on climate change stressors.

VULNERABILITY STUDIES ON CHONDRICHTHYANS FROM CLIMATE CHANGE STRESSORS IN MEXICO

The studies developed in Mexico are those undertaken by Garcés-García (2020). She developed an investigation in which she assessed the vulnerability to climate change stressors of 106 species of sharks, rays, and chimaeras that are distributed in western Mexico. The oceanographic conditions such as water temperature are important for some sharks and rays, though it is unclear to what extent. For example, sea surface temperatures are cold close to Baja California in Mexico but warmer waters are in southern Mexico, close to the border with Guatemala. This factor can influence where certain species are found. But, those animals with a large distribution range can tolerate different water temperatures, so this information is useful for scientists who investigate how adaptable the animal is to new conditions or new water temperatures. On the other hand, some regions are difficult to study because information is incomplete. Examples of regions that are difficult to study are the inside part and entrance of the Gulf of California as well as the rest of the Mexican Pacific (Figure 2.8).

Based on this research, Garcés-García concluded that the Gulf of California region is an important area for chondrichthyan species that are currently coping with climate change stressors because the

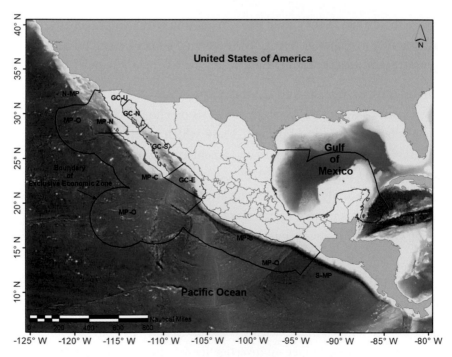

Figure 2.8 *Gulf of California (GoC) and its internal and external subregions for evaluation of distributional flexibility of the chondrichthyan species.*

oceanographic conditions are more stable inside the Gulf of California and are highly resilient.

TEMPERATURE RANGES FOR THE STUDY AREA

Furthermore, Garcés-García et al. (2021) developed a study for the Gulf of California in which six species of sharks, that are included in the Convention on International Trade in Endangered Species of Wild Fauna and Flora (CITES), can be potentially affected by climate change stressors, especially changes in sea surface temperature (Figure 2.9).

In this investigation the authors evaluated the following sharks: *Alopias pelagicus* (pelagic thresher), *A. superciliosus* (bigeye thresher), *A. vulpinus* (common thresher), *Carcharhinus falciformis* (silky shark), *Isurus oxyrinchus* (shortfin mako) and *I. paucus* (longfin mako). These sharks are species that produce few offspring, of which historical population declines have been documented due to the international trade in fins and meat (CITES 2016). In this investigation, some important stressors

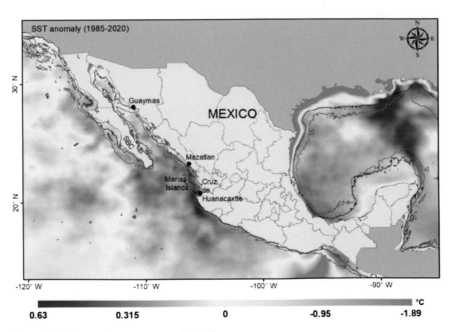

Figure 2.9 *Temperature ranges for the study area.*

that were studied for their effect on the six species of sharks were 1) changes in sea surface temperature, 2) changes in atmospheric pressure, 3) upwelling index (which represents the speed of water movement from the sea bottom to the surface), 4) changes in the amount of dissolved oxygen in the water, 4) changes in the amount of chlorophyll-*a* (which is a pigment that gives terrestrial and aquatic plants the green color), 5) changes in salinity, and 6) pH (which shows how acidic the ocean is). The study projected parameters for the year 2100, based on some greenhouse emissions scenarios (categorized as low, medium, and high emissions) in order to evaluate what the environmental conditions in the study area would be and how the six sharks could be affected (the vulnerability was expressed as low, medium, and high; however, some mathematical background was applied which is not included in this section). Overall, the authors concluded that the six species are under low vulnerability, which means the six sharks will not be severely affected. Nevertheless, the authors show that some shark attributes, like habitat dependence, can increase vulnerability. An example is the pelagic thresher shark which needs nearshore nursery areas, which are increasingly affected by climate change, for pupping (Salomón-Aguilar et al., 2009).

VULNERABILITY STUDIES IN OTHER PARTS OF THE WORLD

Another valuable investigation related to which species of chondrichthyans are affected by climate change was done by Chin et al. (2010) for the Great Barrier Reef (GBR) in Australia. The authors evaluated more than 130 species of sharks, skates, and rays. They developed a study and considered many factors such as water temperature, ocean acidification, freshwater input, ocean circulation (how strong or weak the water in the ocean moves), sea level rise, severe weather and light. The authors found that for the GBR, freshwater/ estuarine and reef associated sharks and rays are most vulnerable to climate change because these areas are more exposed to changes in the factors mentioned above. Other studies have focused on commercial sharks. For instance, Birkmanis et al. (2020) evaluated the requiem and mackerel sharks. They used shark occurrence records collected by commercial fisheries within the Australian continental region and they considered the period 2050–2099. The researchers were focused on two main factors such as sea surface temperature and turbidity to develop a climate change model using special computing programs. They concluded that a shift in good habitat conditions for both requiem and mackerel sharks within the study area is likely to occur in the period 2050–2099.

Another remarkable study was developed by Walker et al. (2021) for Southern Australia. The authors built for the first time a framework that involves fishing and climate change factors. They studied more than 130 sharks and rays that are found in Southern Australia.

Dr Terry Walker, who was the leader of this investigation, and his coauthors worked on a method to include many factors that can affect sharks and rays. Their approach is so exciting because it allows scientists to evaluate quickly which animals need urgent protection, and which species could be in danger by the year 2100. The study used three hypothetical different climate change scenarios known as low, medium, and high as well as eight climate change factors including rising water temperature (at 75 m depth), rising sea level, changing rainfall and freshwater runoff, increasing storm frequency and intensity, changing marine currents and upwelling index, increasing UV light radiation, decreasing oxygen in the water, and increasing ocean acidity. These findings are therefore extremely important to prepare in advance. The authors concluded that the three climate change scenarios that are mentioned above indicate rising vulnerability for the 130 species. But there is one type of skate named the "maugean skate" which was evaluated at medium vulnerability. This is because the skate is found in inshore waters where the severity of climate change effects is more remarkable. So, this gives us an idea of the difficulties that some species in inshore areas will face. By now these studies are available where many species can be evaluated. However, more climate change studies on elasmobranch species are needed. It is known that sharks are vulnerable to different factors, but it is unknown which species are at most vulnerable in many other parts of the world or which type of factors are more important. Therefore, it is urgent to continue applying these types of studies to protect them and make better conservation decisions.

UNDERSTANDING SHARK ECOLOGY THROUGH OBSERVATIONAL STUDIES

UNDERSTANDING SHARK ECOLOGY THROUGH OBSERVATIONAL STUDIES

Florida is home to many diverse species of sharks, and shark tourism has increased in popularity in recent years. Many travel from around the world to dive with tiger, lemon, silky, and great hammerhead sharks, all of which are near threatened or endangered according to the International Union on the Conservation of Nature (IUCN).

The Southeast Florida coastline is rich in marine biodiversity, in part because of its close proximity to the Gulf Stream. Several individual shark species receive a lot of research attention but shark assemblages, different shark species that share the same area, are less studied. This approach can help develop management frameworks for multi-species sustainable fisheries and help scientists understand what changes the shark assemblages go through over time, what may cause those changes, and whether there are ways to mitigate any negative impacts. American Shark Conservancy (ASC) is a non-profit organization founded by lead scientist Hannah Medd, MSc in Jupiter, Florida. ASC's project, "Florida Shark Survey," is a long-term monitoring program established in 2015, with the goal of understanding the distribution of different shark species and the impact of anthropogenic actions on local shark populations. It is the first project of its kind to use a combination of methods to record the diversity, abundance, and behavior of a variety of shark species in Florida. ASC scientists use non-invasive, in-water biological and behavioral data collection techniques. One such technique is to measure precise length measurements with paired-laser photogrammetry. ASC uses Underwater Visual Census (UVC) methods to positively identify the species of sharks through cameras, determine the sex, and record other details like the number and types of parasites, damage from and the retention of fishing gear, as well as any interesting behaviors. The results and images from this study will be used to determine shifts in shark species or abundance over time, helping to determine what may cause these shifts, and how to reduce the negative impacts. These data can also help identify behavior between shark species in Florida. The survey data have been presented at a scientific conference and provided to federal fisheries managers to help designate certain south Florida areas as an HAPC (Habitat Area of Particular Concern) for lemon sharks (*Negaprion brevirostris*). The results are also easily used for public outreach to encourage the community to better understand the vulnerable species off their own coast.

The dive community and the general public contributes to the data collected, because it is simple to participate in, and because it is important to involve the community and stakeholders in research and science. A few popular "locals" have been observed by the survey divers for several years. For example, Snooty the lemon shark (*Negaprion brevirostris*) is known as the celebrity shark of Jupiter, Florida, and can be identified by her broken back tail fin and her teeth sticking out. Through monitoring the same local sharks for changes in behavior and how they mature, interesting behavioral findings have been noted for several shark species over the last two to three years of the Florida Shark Survey. For example, lemons are mostly coastal in Florida and like to hang out around wrecks where people dive. At deeper dive locations, lemon sharks (*Negaprion brevirostris*), tend to position themselves at the top of the group, taking the bait and pushing all the surrounding pelagic, deeper water species out.

Another example is Patrick, a male tiger shark (*Galeocerdo cuvier*) estimated to be 5–6 years old when identified over three years ago. He was initially 7ft when first measured, now has grown to 10ft, and his diet expanded to include barracuda. Survey data supports the migratory behaviors of tiger sharks that the American Shark Conservancy (unpublished data) has observed and that certain individuals return to the same location during the same time period each year. In 2021, Patrick left at the end of March, came back the first week of April, and left again the first week of July, suggesting that some male tiger sharks tend to not travel far.

Additional funding through the National Geographic Young Explorer grant is allowing for the opportunity to expand the shark survey program to include students (ages 17–23, of all genders and identities) from underrepresented communities where they are most in conflict with shark conservation. Through collaborative community science participation with local students, the shark survey program will continue to provide important information to develop management frameworks for multi-species sustainable fisheries and help scientists understand the cause and effect of change in shark assemblages over time (Figures 2.10–2.13, 2.14, 2.15, 2.16, 2.17).

CONCLUSION

While a variety of techniques can be used to study the ecology and evolution of shark species, new and innovative techniques are constantly being developed. Now that advanced technologies such as

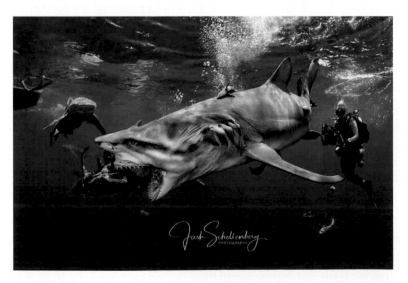

Figure 2.10 Bull shark eating bait. Photo credit Josh Scheldenberg.

Figure 2.11 *Divers swimming with lemon sharks near a wreck in Jupiter, FL. Photo credit Applecorps photography.*

Figure 2.12 *Tiger shark swimming upwards. Photo credit Josh Scheldenberg.*

CT scans, computer software, and molecular techniques exist, scientists can collect and analyze massive amounts of data and study sharks in minimally invasive ways. However, all methods have their pros and cons and just because new technology exists does not mean it is always the best solution. Sometimes, there is no substitute for traditional science

Figure 2.13 Photo of Hannah Medd, lead scientist for American Shark Conservancy, conducting a survey.

Figure 2.14 National Geographic Shark Survey research trip #1 with students (CH2-Understanding shark ecology through observational studies). Photo credit, Cassandra Scott.

Figure 2.15 *Setting up gear for Shark Survey with ASC Scientists (Gretchen Kruizenga, Hannah Medd, Kaitlyn Issacson), and volunteer (Sabrina Wilner). Photo credit: Cassandra Scott.*

Figure 2.16 *Student meeting with Deborah Santos de Azevedo Menna to discuss survey data collected (CH2-Understanding shark ecology through observational studies). Photo credit, Gretchen Kruizenga.*

Figure 2.17 *Student taking length measurements of shark using laser rig (CH2-Understanding shark ecology through observational studies). Photo credit, Cassandra Scott.*

techniques and keen observation. While fancy equipment and complex techniques can be great tools to answer some ecological and evolutionary questions, many can be studied in simpler, cheaper, and more community-driven ways with widely available tools and a little creativity.

REFERENCES

Birkmanis, C. A., Freer, J. J., Simmons, L. W., Partridge, J. C., & Sequeira, A. M. M. (2020). Future distribution of suitable habitat for pelagic sharks in Australia under climate change models. *Frontiers in Marine Science, 7*, 570. https://doi.org/10.3389/fmars.2020.00570

Bonfil, R., Meÿer, M., Scholl, M. C., Johnson, R., O'Brien, S., Oosthuizen, H., Swanson, S., Kotze, D., & Paterson, M. (2005). Transoceanic migration, spatial dynamics, and population linkages of white sharks. *Science, 310*(5745), 100–103. http://www.jstor.org/stable/3842870

Carrier, J. C., Heithaus, M. R., & Simpfendorfer, C. A. (Eds.). (2019). *Shark research: Emerging technologies and applications for the field and laboratory.* Boca Raton, FL: CRC Press.

Cavalcanti, M. J. (2007). A phylogenetic supertree of the hammerhead sharks (Carcharhiniformes: Sphyrnidae). *Zoological Studies, 46,* 6–11.

Charlesworth, B. (2009). Effective population size and patterns of molecular evolution and variation. *Nature Reviews Genetics, 10*(3), 195–205.

Chin, A., Kyne, P. M., Walker, T. I., & McAuley, R. B. (2010). An integrated risk assessment for climate change: Analysing the vulnerability of sharks and rays on Australia's great barrier reef. *Global Change Biology, 16*(7), 1936–1953. https://doi.org/10.1111/j.1365-2486.2009.02128.x

Chin, A., Tobin, A., Simpfendorfer, C., & Heupel, M. (2012). Reef sharks and inshore habitats: Patterns of occurrence and implications for vulnerability. *Marine Ecology – Progress Series, 460,* 115–125.

Chivian, E., & Bernstein, A. (2010). How our health depends on biodiversity. In *Center for health and the global environment* (pp. 5–22). Boston, MA: United Nations.

Compagno, L. J. V. (1988). *Sharks of the order Carcharhiniformes.* Princeton, NJ: Princeton University Press, pp. 13, 357–361.

Compagno, L. J. V. (1990). Alternative life-history styles of cartilaginous fishes in time and space. *Environmental Biology of Fishes, 28*(1–4), 33–75.

Dudgeon, C. L., Blower, D. C., Broderick, D., Giles, J. L., Holmes, B. J., Kashiwagi, T., … Ovenden, J. R. (2012). A review of the application of molecular genetics for fisheries management and conservation of sharks and rays. *Journal of Fish Biology, 80*(5), 1789–1843.

Fields, A. T., Abercrombie, D. L., Eng, R., Feldheim, K., & Chapman, D. D. (2015). A novel mini-DNA barcoding assay to identify processed fins from internationally protected shark species. *PLOS ONE, 10*(2), e0114844.

Garcés-García, K. C. (2020). Effects of fishing and climate change on the chondrichthyan species in the Gulf of California region. Doctoral Thesis. The University of Melbourne, Australia, 263 p.

Garcés-García, K. C., Day, R. W., Walker, T. I., Tovar-Avila, J., Castillo-Géniz, J., & Godinez-Padillas, C. J. (2021). Tiburones incluidos en los apéndices de la cites, y el efecto potencial del cambio climático. El Golfo de California como caso de estudio. In *Tiburones Mexicanos en CITES* (Vol. II, pp. 79–90). Coyoacán, Mexico: Instituto Nacional de Pesca y Acuacultura.

Gibbs, L., & Warren, A. (2014). Killing sharks: Cultures and politics of encounter and the sea. *Australian Geographer, 45*(2), 101–107. https://doi.org /10.1080/00049182.2014.899023

Grogan, E. D., & Lund, R. (2004). The origin and relationships of early Chondrichthyans. In J. C. Carrier, J. A. Musick, & M. R. Heithaus (Eds.), *Biology of sharks and their relatives* (pp. 3–32). Boca Raton, FL: CRC Press LLC.

Hey, J., Waples, R. S., Arnold, M. L., Butlin, R. K., & Harrison, R. G. (2003). Understanding and confronting species uncertainty in biology and conservation. *Trends in Ecology and Evolution, 18*(11), 597–603.

Hillis, D. M. (1998). Taxonomic sampling, phylogenetic accuracy, and investigator bias. *Systems Biology, 47*(1), 3–8.

Hughes, A. R., Inouye, B. D., Johnson, M. T., Underwood, N., & Vellend, M. (2008). Ecological consequences of genetic diversity. *Ecology Letters, 11*(6), 609–623.

IPCC. (2007). *Climate change 2007: The physical science basis.* Cambridge University Press. https://doi.org/10.1256/wea.58.04.

IPCC. (2014). Summary for policymakers. In *Climate change 2014: Synthesis report. Contribution of working groups I, II and III to the fifth assessment report of the intergovernmental panel on climate change.* https://doi.org/10 .1017/CBO9781107415324.

Joel, L. (2019). Jaws: The never-ending story. *New Scientist, 242*(3236), 43–45.

Kajiura, S. M., Forni, J. B., & Summers, A. P. (2003). Maneuvering in juvenile carcharhinid and sphyrnid sharks: The role of the hammerhead shark cephalofoil. *Zoology, 106*(1), 19–28.

Knip, D. M., Heupel, M. R., & Simpfendorfer, C. A. (2010). Sharks in nearshore environments: models, importance, and consequences. *Marine Ecology Progress Series, 402*, 1–11.

Kuguru, G., Gennari, E., Wintner, S., Dicken, M. L., Klein, J. D., Rhode, C., & Bester-van der Merwe, A. E. (2019). Spatio-temporal genetic variation of juvenile smooth hammerhead sharks in South Africa. *Marine Biology Research, 15*(10), 568–579.

Lim, D. D., Motta, P., Mara, K., & Martin, A. P. (2010). Phylogeny of hammerhead sharks (Family Sphyrnidae) inferred from mitochondrial and nuclear genes. *Molecular Phylogenetics and Evolution, 55*(2), 572–579.

Luikart, G., Ryman, N., Tallmon, D. A., Schwartz, M. K., & Allendorf, F. W. (2010). Estimation of census and effective population sizes: The increasing usefulness of DNA-based approaches. *Conservation Genetics, 11*(2), 355–373.

Martínez-Candelas, I. A., Pérez-Jiménez, J. C., Espinoza-Tenorio, A., McClenachan, L., & Méndez-Loeza, I. (2020). Use of historical data to assess changes in the vulnerability of sharks. *Fisheries Research*, 226, 105526.

Naylor, G. J. P. (1992). The phylogenetic relationships among requiemand hammerhead sharks: Inferring phylogeny when thousands of equally most parsimonious trees result. *Cladistics*, 8(4), 295–318.

Naylor, G. J., Caira, J. N., Jensen, K., Rosana, K. A., Straube, N., & Lakner, C. (2012). Elasmobranch phylogeny: A mitochondrial estimate based on 595 species. *Biology of Sharks and Their Relatives*, 2, 31–56.

Ovenden, J. R., Leigh, G. M., Blower, D. C., Jones, A. T., Moore, A., Bustamante, C., ... Dudgeon, C. L. (2016). Can estimates of genetic effective population size contribute to fisheries stock assessments? *Journal of Fish Biology*, 89(6), 2505–2518.

Padial, J. M., Miralles, A., De la Riva, I., & Vences, M. (2010). The integrative future of taxonomy. *Frontiers in Zoology*, 7(1), 1–14.

Pardini, A. T., Jones, C. S., Noble, L. R., Kreiser, B., Malcolm, H., Bruce, B. D., ... Martin, A. P. (2001). Sex-biased dispersal of great white sharks. *Nature*, 412, 139–140.

Pollock, D. D., Zwickl, D. J., McGuire, J. A., & Hillis, D. M. (2002). Increased taxon sampling is advantageous for phylogenetic inference. *Systems Biology*, 51(4), 664.

Portnoy, D. S., & Heist, E. J. (2012). Molecular markers: Progress and prospects for understanding reproductive ecology in elasmobranchs. *Journal of Fish Biology*, 80(5), 1120–1140.

Quattro, J. M., Stoner, D. S., Driggers, W. B., Anderson, C. A., Priede, K. A., Hoppmann, E. C., ... Grady, J. M. (2006). Genetic evidence of cryptic speciation within hammerhead sharksb (genus Sphyrna). *Marine Biology*, 148(5), 1143–1155.

Rannala, B., Huelsenbeck, J. P., Yang, Z., & Nielsen, R. (1998). Taxon sampling and the accuracy of large phylogenies. *Systems Biology*, 47(4), 702–710.

Roff, G., Doropoulos, C., Rogers, A., Bozec, Y. M., Krueck, N. C., Aurellado, E., ... Mumby, P. J. (2016). The ecological role of sharks on coral reefs. *Trends in Ecology and Evolution*, 31(5), 395–407.

Salomón-Aguilar, C. A., Villavicencio-Garayzar, C. J., & Reyes-Bonilla, H. (2009). Zonas y temporadas de reproducción y crianza de tiburones en el Golfo de California: Estrategia para su conservación y manejo pesquero. *Ciencias Marinas*, 35(4), 369–388. https://doi.org/10.7773/cm.v35i4.1435.

Stefanovic, S., Rice, D. W., & Palmer, J. D. (2004). Long branch attraction, taxon sampling, and the earliest angiosperms: Amborella or monocots? *BMC Evolutionary Biology*, *4*, 35.

Timmerman, P. (1981). Vulnerability, resilience and the collapse of society. In P. Timmerman (Ed.), *Environmental monograph* (pp. 1–38). Toronto: University of Toronto.

Walker, T. I., Day, R. W., Awruch, C. A., Bell, J. D., Braccini, J. M., Dapp, D. R., Finotto, L., Frick, L. H., Garcés-García, K. C., Guida, L., Huveneers, C., Martins, C., Rochowski, B., Tovar-Ávila, J., Trinnie, F. I., & Reina, R. D. (2021). Ecological vulnerability of the chondrichthyan fauna of southern Australia to the stressors of climate change, fishing and other anthropogenic hazards. *Fish and Fisheries*, *22*(5). https://doi.org/10.1111/faf.12571

Ward, R. D., Zemlak, T. S., Innes, B. H., Last, P. R., & Hebert, P. D. (2005). DNA barcoding Australia's fish species. *Philosophical Transactions of the Royal Society of London Series B: Biological Sciences*, *360*(1462), 1847–1857.

Ward-Paige, C. A., & Worm, B. (2017). Global evaluation of shark sanctuaries. *Global Environmental Change*, *47*, 174–189.

Zwickl, D. J., & Hillis, D. M. (2002). Increased taxon sampling greatly reduces phylogenetic error. *Systems Biology*, *51*(4), 588–598.

3

Fields of elasmobranch anatomy and physiology

Written by Lauren Eve Simonitis with contributions
from Miasara Andrew, Tatyana Brewer-Tinsley,
Aubree Jones, Peyton Thomas, Sabrina Van Eyck,
and Amani Webber-Schultz

DOI: 10.1201/9781003260370-3

Dr Lauren Eve Simonitis

 I am a Cuban-American aquatic sensory biologist born and raised in Miami, FL, USA. I grew up around, on, and in the water snorkeling, SCUBA diving, and paddle boarding. When I learned in my high school biology class that I could have a career studying the ocean, I started working toward becoming a marine biologist. I went to the University of Miami where I double majored in Marine Science and Biology. I delved right into research, where I developed a love of neuroscience in Dr Lynne Fieber's lab studying the nervous system of California sea hares and how different neurotransmitters affect the growth and lifespan of neurons. I also participated in the National Science Foundation's Research Experience for Undergraduates program at the University of Alaska-Anchorage with Dr Jennifer Burns. In Dr Burns' lab, I explored how harp and hooded seal muscles change throughout their switch from land-dependent babies to expert swimmers. This research sparked my interest in adaptations to life in water. For graduate school, I set out to combine both of my interests, focusing on how aquatic animals sense their environment. After graduating in 2015, I enrolled in the Marine Biology PhD program at Texas A&M University at Galveston working with Dr Christopher Marshall in his Ecomorphology and Comparative Physiology lab.

As an aquatic sensory biologist, I focus both on how animals sense their environment and how they interact with the senses of other animals. Sharks provided an exciting animal to study in this capacity, due to their wide range of sensory capabilities! For my dissertation, I focused on how sharks use their sense of smell and how prey species can weaponize chemicals to attack this sense. Specifically, I looked at how prey animals use ink to defend themselves against sharks. I studied ink from a variety of organisms like California sea hares, common cuttlefish, and these strange inking whales—pygmy sperm whales. Ink attacks predators in a couple of ways: 1) it is dark and acts as a smoke screen, 2) it is sticky, and 3) it has nasty-smelling chemicals in it. To test how this worked on sharks, I would expose swimming bonnethead sharks to ink and record their behavioral reactions (spoiler alert: they did not like it). After seeing these reactions, I decided to look into the nose of the shark—since we think this is the sense that the ink is attacking. I was shocked to see how little we know about shark noses! So now, I focus on how sharks use their sense of smell. I look at both water flow and sensory structures distribution inside differently shaped shark noses. I also look at how water flows within these shark noses and how something sticky (like ink) can affect that flow. I also am looking at the physiology behind the shark sense of smell, using electro-olfactograms to record the actual responses of the shark's nose in response to different smells.

Miasara Andrew

 I was born in New York and have always been in love with the ocean. Sharks are where my fascination with the ocean started. I was always so intrigued by these animals because of how much the people surrounding me feared them. I grew up around the suburbs and the cities of New York which kept me from opportunities to explore the ocean. So I moved to Florida a few years ago to pursue Marine Biology. I received my Bachelor's degree in Biology with a concentration in Marine Science from Eckerd College in 2019. I am currently a graduate student at the University of Miami's Rosenstiel School of Marine and Atmospheric Science. I spend a lot of time in the ocean through my hobbies of SCUBA diving, freediving, and underwater photography. I recently spent some time living out at sea by participating in a research cruise and working as a research assistant in NOAA's Ocean Acidification program. This opportunity has allowed me to explore the ocean in a different way.

My interest in shark research developed during my BSc where I had the opportunity to assist in shark research at the Bimini Biological Field Station. I became fascinated by *electroreception* and how sharks are able to interact with their environment through the use of their electroreceptive organs. I am interested in the physiology and morphology of sharks' sensory systems. I am particularly interested in how sharks use these systems to interact with their environment. In the summer of 2019 I spent some time researching sharks in South Africa which allowed me to learn more about the endemic species in South Africa. My research included monitoring shark biodiversity in South African waters and measuring the healing rates of wounds on sharks. I am currently working on my MSc thesis project focusing on the effects of ocean acidification on sharks—specifically their migration patterns based on dissolved oxygen concentrations. In the future I am looking to combine this with shark sensory systems and evolution to further understand these animals and apply this to future conservation.

Tatyana Brewer-Tinsley

My name is Tatyana Brewer-Tinsley and I am a Black American woman who has always dreamed of educating the public about the beauty of our oceans. Hailing from the Bronx in the concrete jungle of New York City, I did not always have the opportunity to explore the seas like I would have liked to. From as young as I could remember, I would often hear about the fears of the unknown surrounding the water and the organisms that inhabit it. Since then, I have made it my mission to help lower this stigma. I often say that I live on land, but my heart belongs to the water (I'd like to think of myself as a sea lion). If I was not swimming, I was talking about the newest animal factoid I learned or watching an ocean documentary on television. As I got older, this drive pushed me to pursue my BSc at Florida International University (FIU). In December of 2020, I graduated with honors with a double major in Marine Biology and Interdisciplinary Studies. While studying at FIU, I signed up for an upper division elective course in physiological ecology that was led by Dr Yannis Papastamatiou. Within this course, Dr Papastamatiou discussed the behavioral and physiological adaptations that aided marine animals in their survival within the aquatic environment. It was because of this elective that I developed a high interest in diving into the world of physiological research within marine mammal and elasmobranch species. Shortly after taking the class, I applied and was selected to intern within the Predator Ecology & Conservation Lab at FIU to assist with a mercury ecotoxicology study of elasmobranch species within Florida waters under Laura García Barcia.

I would like to pursue my MSc and potentially my PhD within the field of physiology. The scope of the specialization is what I need to narrow down at this time. As a whole, I am interested in researching stress physiology, conservation physiology, comparative physiology, and/or the impact of fear ecology on physiological processes. By being able to participate in this research, I would like to assist with outreach and conservation programs, work within a research laboratory, and potentially teach at a university.

Aubree Jones

I was born in Tulsa, Oklahoma and grew up fishing the lakes and rivers of Oklahoma with my dad, which is largely how I became interested in the natural world. I fell in love with the ocean through the glass of the Oklahoma Aquarium, where I discovered a whole world of amazing fish and other marine life beyond the lakes of Oklahoma. After volunteering at the aquarium in high school, I decided to pursue a degree in Marine Biology. I left my home state and received my BSc in Marine Biology with a minor in Chemistry in 2017 from Texas A&M University at Galveston. I graduated with honors and completed an Honors thesis with Dr. Christopher Marshall. After contemplating what I wanted for my future career, I decided to pursue a graduate degree in 2018 at the University of Rhode Island working with Dr. Jacqueline Webb.

My research interest in sensory biology started at Texas A&M. During my undergraduate degree, I began volunteering on a project in Dr. Marshall's lab on the sensory biology of harp seal whiskers, which was the MSc project for Erin Mattson, MSc. Shortly after, I started my own undergraduate thesis on the sensory biology of harbor seal whiskers, which won multiple awards and was published in 2018. From my work on seal whiskers, I became fascinated by how sensory systems work. I decided to continue studying sensory biology, but for my PhD I have turned my attention to fish. I now study the *lateral line* system in fish, which similar to seal whiskers functions to detect water flows. I study the development and sensory-mediated behavior of this flow sensing system in minnows and trout. My current work strives to achieve two goals: (1) I am looking at the development of adaptive phenotypes (the appearance of a structure/system) to try and understand how developmental processes produce unique or unusual phenotypes. (2) I also focus on how alterations to the environment by human-caused global climate change may affect the development and function of sensory systems. I am researching how increased temperatures affect the development of the lateral line system in brook trout, an iconic coldwater species. I am also examining how changing the sediment of the substrate impacts the sensory capabilities of minnows that pick prey items out of the bottom substrate.

It has been a joy that my current work brings me back to the lakes, rivers, and streams that are familiar to my experiences growing up fishing in Oklahoma. The ocean is a charismatic system and has been the focus of much discourse regarding the effects of climate change; but growing up around freshwater systems and understanding their importance first-hand, I strive to bring awareness to the threats facing freshwater ecosystems. Even today I enjoy getting out on the water as much as possible and enjoy paddle boarding, kayaking, fishing, hiking, yoga, and exploring the outdoors with my ultimate adventure partner, my dog Winston.

Peyton Thomas

I am a black fish physiologist with a love for marine systems. I grew up in Atlanta, Georgia, where I found my interest in aquatic systems through venturing out into the local rivers and streams. Throughout my early school years I knew I wanted to work with animal systems but didn't find out my true passion for aquatic animals until college. I attended college

at Baylor University where I began as a Pre-Veterinary student but transitioned to a path in Environmental Science because I wanted to focus on specific environmental issues like pollution impacts to the ecology of various organisms. With this transition came an opportunity to join a lab that focused on microplastic pollution in both freshwater and marine fish in Texas, and this is where I got my first real glimpse of the marine environment. Growing up in suburban Georgia didn't allow many opportunities to get to the coast beyond the Georgia Aquarium, which wasn't operational until my middle school years. My time as an undergraduate researcher solidified that I was indeed curious about the marine environment and wanted to better understand the impacts to organisms on a smaller scale, beginning with their basic functioning under various types of environmental stress. This led me to initially pursue an MSc in Marine Science at the University of North Carolina at Wilmington, where I worked on a project focused on the impacts of various anti-fouling treatments on the growth and viability of Eastern oysters impacted by heavy fouling events of an invasive tunicate. With the events of Hurricane Florence in 2018 came the abrupt end to that project, but the light of a new opportunity to delve into the world of elasmobranchs, the fishes I was most curious about. This led me to where I am today, my dissertation focused on elasmobranch muscle morphology and oxidative stress under elevated temperature stress. I'm broadly interested in elasmobranch physiology, but more specifically, the energetic demands of various environmental stressors on elasmobranchs and the impacts on their survival. Often, we view impacts of environmental stress from a top-down perspective, not fully understanding changes occurring on the molecular and cellular levels. I'm also interested in environmental policy and making science more accessible to the general public through community science opportunities.

I would like to pursue my MSc and potentially my PhD within the field of physiology. The scope of the specialization is what I need to narrow down at this time. As a whole, I am interested in researching stress physiology, conservation physiology, comparative physiology, and/or the impact of fear ecology on physiological processes. By being able to participate in this research, I would like to assist with outreach and conservation programs, work within a research laboratory, and potentially teach at a university.

Sabrina Van Eyck

I was born in Brussels, Belgium and moved to the Democratic Republic of Congo shortly afterward, where I lived with my family for almost three years. We then moved to the San Francisco Bay Area in California, and I have been here ever since. I grew up going to the beach quite often with my family, and very early on developed a love for the ocean and science. I spent much of my childhood either at the beach or watching documentaries about the ocean. I received a toy microscope one year for my birthday and that really sparked my interest in science. I was always so amazed by the things I could see under the lens. In high school, I took a course on marine biology and that helped me decide that I was going to go into that field. After watching the documentary Sharkwater, I decided that I really wanted to help sharks and be a voice for them. I ended up getting my BSc in Marine Biology from San Jose State University. During this time, I worked in Dr Scott Shaffer's Aviary Lab, where I studied the diet of albatross in different locations. Most of my courses were on marine biology, zoology, and the physiology of marine mammals. I obtained my PADI Advanced Open Water Diver certification while at school as well. I was fortunate enough to also do a couple of internships at the Marine Science Institute in Redwood City, CA. I assisted the instructors with running stations for the field trips that happened on the R.V. Robert G. Brownlee. I also helped feed and maintain the aquarium onsite, as well as cultivate algae.

Currently, I am a Specialty Programs guide at the California Academy of Sciences, where I lead behind-the-scenes tours and help create educational programs tailored to families. Lately, I have also been helping spread the word about the Academy's current initiatives for helping our planet. I also talk about shark conservation in the new traveling exhibit.

I am very interested in shark science and conservation, and I hope to educate the public on their importance. I truly enjoy teaching science and inspiring young minds. My goal is to inspire people of all ages to care more about our oceans and these apex predators. In my free time, I very much enjoy traveling, SCUBA diving, snorkeling, and just being in and around the ocean in general.

Amani Webber-Schultz

I am a biracial shark scientist. I'm also a co-founder and the Chief Financial Officer of Minorities in Shark Sciences Inc. I was born in Washington DC, USA. Shortly after, my two moms moved to the Bay Area in California where I grew up. I was exposed to the ocean from a young age through family vacations, trips to the Monterey Bay Aquarium, the San Francisco Bay, and the Pacific Ocean near my house. Like many young children, I jumped between what career I wanted to pursue but somehow always ended up back at marine biology. The summer before my senior year of high school, I participated in a program removing invasive species from coral reefs which solidified my desire to pursue marine biology as a major in college. I attended Rutgers University for college, majoring in marine biology and minoring in fishery science. During my BSc I worked in a paleoceanography lab where I sorted foraminifera out of deep sea sediment samples for climate studies. I also created 3-D models of remoras for morphological studies on their soft tissues. My love of sharks came from participating in shark field research with Field School in Miami, USA (www.getintothefield .org). It was there that I discovered how much joy field research, being on boats, and interacting with sharks bring me. In 2020, I co-founded Minorities in Shark Sciences Inc., an organization dedicated to creating a welcoming space and opportunities for women of color to pursue shark science as a career. I am very passionate about supporting BIPOC in STEM since those fields have historically excluded them. I also enjoy mentoring and teaching since I am lucky enough to have amazing mentors in my life and want to pass on that gift to others.

In 2021, I started as a PhD student at the New Jersey Institute of Technology. I am interested in shark morphology, particularly their scales, termed *dermal denticles*, which have super cool microstructures and are definitely underrated. I hope to study how their skin benefits them in water and research the evolution and development of the different shapes and sizes of dermal denticles. I am currently working on remoras, also known as suckerfish. In addition to graduate school, I also co-host a podcast with Meghan Holst called the Sharkpedia Podcast where we break down elasmobranch-related scientific papers so that non-scientists and scientists alike may better understand them. In my free time, I rock climb, hang out with friends, and seek out really good food.

KEY TERMS

action potential: an electrical signal sent from a nerve to communicate information to the brain

ampullae of Lorenzini: sensory organs that detect electric fields found in sharks, skates, and rays

dermal denticles: teeth-like scales found exclusively in sharks, skates, and rays

electroreception: sensing electric fields

lateral line: the sensory system that detects local water flows in sharks, skates, and rays

magnetoreception: detection of earth's geomagnetic waves

mechanosensation: the sense of touch—detection of pressure or vibration; in aquatic vertebrates this sense is detected via water flows

metabolism: combustion of carbohydrates, amino acids, and fats

myoglobin: molecule which stores oxygen in the muscle

olfaction: the sense of smell—detection of chemical stimuli by the nose

regional endothermy: maintaining higher temperatures than the outside environment in a specific area of the body

INTRODUCTION

Prior to our current sixth mass extinction event, largely caused by human interactions with the environment, sharks managed to survive five previous mass extinction events. Their success is a testament to the incredible adaptations that they have developed over the millions of years they have inhabited the oceans. Studying their anatomy and morphology (how their bodies look) along with their physiology (how their bodies work) is one of the best avenues scientists have to get a glimpse into how a shark works and succeeds in its environment. Anatomy and physiology are linked via the "form function paradigm." Essentially, this means that how a part of their body looks will tell us how it works and vice versa.

Biological success revolves around two things—food and sexual reproduction. The ability of an animal to fuel its body and pass on its genes are the two main ingredients in the recipe for the "how to outlive the dinosaurs" cocktail. This chapter will focus on the former, detailing how a shark's different body parts and systems work to successfully complete a predation event. Sharks are found at all different places in the food web ranging from high up mega-predators to chill plankton eaters and even an invertebrate and seagrass salad connoisseur. Despite differences in diet, a predation event sticks to a general formula.

The first step of a predation event is the encounter, where sharks can use their sensory capabilities to detect their food item. At this stage, sharks use both long range senses such as hearing and *olfaction*, as well as short range senses such as *mechanosensation*, and *electroreception*. Once the prey has been detected, the attack is initiated where sharks use their powerful swimming muscles to close in on their target. Next, sharks move into the capture phase, using their mouths to get control of their prey item. Finally, the predation event ends with ingestion as the shark moves the prey item from the mouth to the rest of the digestive system. This chapter delves into each step of the predation event, highlighting the anatomy and physiology that makes a shark a successful predator.

THE ENCOUNTER PT. 1—LONG RANGE SENSORY MORPHOLOGY AND PHYSIOLOGY

A predation event starts off with a bang: a shark perceives its food source by employing a suite of impressive sensory systems. These senses can be divided into two categories: long range and short range, which occur on the scales of tens of meters in distance to within one body length or less, respectively. Early encounters employ the use of the long range senses, giving sharks the ability to lock in on a food source and move in closer. *Magnetoreception* is the ability of a shark to sense the earth's magnetic field, allowing them to navigate across ocean basins and return to feeding grounds. In this section, the authors will delve into the other two long range sensory systems: hearing and olfaction.

HEARING

A speargun fires. An injured fish thrashes in the water. Meters away, a shark hears the call of food and changes the direction of its swimming. The sound waves produced by the spear gun and the thrashing fish travel through the water and reach the shark's ears. For decades, scientists have strived to understand how sharks, skates, and rays hear underwater. What kinds of sounds they are attracted to? What sounds do they avoid?

A sound is a vibration that produces sound waves that travel through the water. These waves can be described using three variables (Figure 3.1): frequency (number of sound waves in a certain time period), wavelength (how long the sound waves are), and amplitude (how much the sound wave displaces the water around it). Scientists typically describe the sensitivity of hearing using frequency. Low frequency sounds have long wavelengths, which means that they travel farther in the water. Sounds typically detected by sharks, like a fish thrashing in the water, are classified as low frequency sounds (below 60 Hz). High frequency sounds have much shorter wavelengths, and do not travel as far, so if the shark or ray can detect this type of sound they must be closer to the sound source.

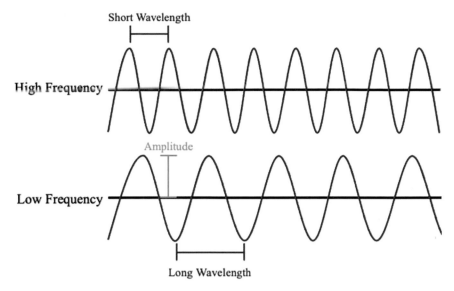

Figure 3.1 *The components of a sound wave. High frequency waves have more sound waves per time period (in red) and shorter wavelengths (in purple). Low frequency waves have less sound waves per time period and longer wavelengths. The amplitude of the wave is shown in blue.*

In seawater, sound travels much faster than in air, so even though low frequency signals can travel far to reach the shark or ray, these signals can still reach the fish rather quickly (Thewissen & Nummela, 2008).

The ear of sharks is contained in the head and looks different from a land animal's ear. This is not surprising since shark ears detect sounds underwater rather than in air (Figure 3.2). There are two major components of the shark ear: the semicircular canals and the otolithic organs. These make up the inner ear. The semicircular canals are three relatively round, looped structures that contain fluid. These canals sense the shark's orientation in its environment and help with balance. This is very similar to humans! When a person sits and spins in an office chair, then suddenly stops—the world feels like it's still spinning. This is because the fluid in their semicircular canal is sloshing around as if they're still going in circles.

Sound waves are detected in a different part of the ear: the chambers located below the semicircular canals (Gardiner et al., 2012). There

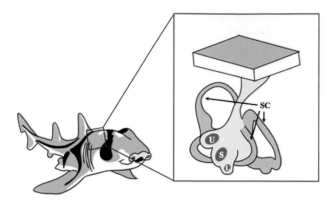

Figure 3.2 *Inside of the ear of a Port Jackson shark (*Heterodontus portusjacksoni*). L – lagena, S – sacculus, U – utriculus, SC – semicircular canals. Modified from Mills et al. 2011.*

are three chambers: the saccule (large), lagena (slightly smaller), and utriculus (smallest). These chambers are filled with fluid and have a sensory surface at their bases. The sensory surface is made up of hair cells, which are named after the hair-like projections (called cilia) poking above the surface of the cells (Figure 3.3). The cilia of the hair cells are covered in a gelatinous matrix (jelly-like substance). Above this sensory surface lie the otoconia, sand-like structures suspended in the jelly. When sound waves come into the ear, they travel into these chambers and vibrate the otoconia which vibrates the jelly underneath them. When the jelly moves, it bends the cilia embedded in this gelatinous matrix. These bending cilia cause the hair cell to send an electric signal (called an *action potential*) which communicates information about this sound to the brain of the shark.

Sharks have an additional unique hearing structure that bony fishes do not have which lets them hear through their skulls! Right under their inner ear near one of the semicircular canals, there is a dip in their skull which houses the macula neglecta. The macula neglecta is connected to the semicircular canal by a duct and is connected to the external skin of the shark by a channel (Fay et al., 1974). Since it is sitting on the skull, sound vibrations travel through the skull and stimulate the hair cells. These hair cells are sensitive to sounds from many different directions, which has led scientists to hypothesize that this organ has a significant

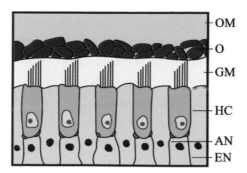

Figure 3.3 A section of the inside of the otolithic organs of the inner ear. The sensory hair cells (blue, HC) make up the sensory epithelium. The hair cells of the inner ear are innervated by afferent (AN) and efferent (EN) nerves. The cilia of the hair cells project into a gelatinous matrix (GM). The otoconia (calcium-carbonate granules, O) sit above the hair cells in the otoconial matrix (OM). When a sound wave causes the otoconia to shift, the gelatinous matrix shifts, which moves the cilia of the hair cells. When the cilia are pushed far enough, the hair cells are activated and they send a signal to the brain of the fish via the afferent nerves. Adapted from Fekete 2003.

role in detecting what directions sounds are coming from. Additionally, the channel leading to the outside of the shark may be a way for sound waves to exit the ear (Corwin, 1981).

Shark and ray ears are most sensitive to sounds similar to the ones their prey make—irregular, low frequency sounds. Most laboratory studies have tested sounds similar to thrashing fish, which elicit attraction responses in sharks (Banner, 1972). These studies have shown that sharks can correctly orient to the direction of a sound with an accuracy to around ten degrees, probably due to the directional sensitivity of the hair cells. Even sharks that were just born (but are free-swimming) respond to sounds that are similar to struggling prey (Banner, 1968). Sharks also respond to noises produced by other fish, which is particularly important on bustling, noisy reefs (Nelson & Johnson, 1972). Sharks and rays themselves are not known to produce sound, so the clues to what kind of sound they are sensitive to come from the sounds of their environment.

Sharks are also capable of learning to respond to specific sounds, which is called sound habituation. One study showed that sharks were able to recognize and respond to music cues. Researchers trained sharks to associate music with a food reward. When researchers later played music, the sharks would show an attraction response, swimming to the location where they receive their food (Nelson et al., 1969; Pouca & Brown, 2018). Sharks can also learn sounds socially, from each other. When one shark is trained to respond to a sound cue, other sharks will learn and begin demonstrating the same response despite never being trained to do so (Nelson & Johnson, 1972). The social aspect of sound habituation gives further evidence to the role sound may play in frenzy feeding behavior. During a frenzy feeding event, one shark's response to a struggling prey item triggers feeding behavior in nearby sharks even if they may not have heard or learned the sound cue. Sharks can even learn non-natural cues. Sharks in areas with lots of spearfishing activity will show an attraction response to the sound of spearguns firing (Nelson & Johnson, 1976).

Not all sounds produce attraction responses though. Sounds can produce flight responses in sharks. Even sounds that may normally be attractive may deter sharks if they suddenly increase drastically in intensity. These sounds make the shark turn and leave, rather than turn and pursue the sound. However, some species respond differently to negative stimuli than others. Silky sharks, for example, would repeatedly withdraw from negative stimuli. However, oceanic white tip sharks would only withdraw once initially, but then they behave neutrally and swim undisturbed towards the negative stimuli. These differences in behavior may be explained by the inner ear of these sharks, but more needs to be done to say for sure (Myrberg, 2001).

Most studies done on shark hearing are carried out in laboratory settings. In laboratories, scientists can control much more of what the shark is sensing. Researchers can limit vision, for example, and control what sounds are played for a shark or ray during the experiment. While this level of control is great to limit outside factors, laboratory studies can hamper the ability of scientists to learn about hearing sensitivity in their natural environment. Sharks in the ocean have to deal with many sounds which may drown each other out. Scientists also don't know how well sharks are able to discriminate between different sounds of varying frequencies and intensities or whether there are seasonal patterns to their hearing sensitivity. Studies on skates have shown that male and female skates have different sensitivity of hearing, which may occur in other sharks and rays as well (Barber et al., 1985). There is extensive debate on which parts of the ear play more significant roles in detecting the direction of sound, and whether sharks have a mechanism for detecting the pressure component of sound (which bony fish sense with a swim bladder—an organ that sharks and rays do not have; van den Burgh and Schuif 1983; Corwin, 1977). Lastly, human activity is increasing the amount of noise in the oceans, with detrimental effects on the hearing of many animals, including sharks and rays. Only a few studies have examined the effects of increased noise on shark behavior, which is a critical area that needs more research (Chapuis et al., 2019). With so many questions to still answer about hearing in sharks, this field is a promising area of research for scientists for decades to come.

OLFACTION

Thanks to movies and TV, sharks have a big reputation for being super smellers. However, despite all this hype, there is not a lot of research in this area. As reviewed by Schluessel et al. (2008), the first studies looking at the shape and function of shark noses began in the early 1900s but ended almost entirely until the 1950s. Spurred by the interest in shark repellents following World War Two, there was a newfound interest in the shark olfactory system and the ways we could agitate it to protect shipwreck victims. Since then, researchers have published on nasal shape, water flow within the nose, size of the parts of the brain associated with smells, the physiology behind smelling, and behavioral responses to smells.

The basic morphology of the olfactory system is standardized among most shark species. Some fish only have two nostrils like humans with water flowing in and out of the same nostril. However, most sharks, like the bonnethead shark (*Sphyrna tiburo;* Figure 3.2), have four nostrils—two on each side of the face. First, water comes in through the incurrent nostril, flowing into the incurrent chamber inside of the olfactory rosette. The olfactory rosette is made up of tissues called the "olfactory lamellae," which are arranged similarly to how dishes are stacked in a dishwasher. These lamellae contain the olfactory receptors neurons. When chemicals in the water bind to the olfactory receptor neurons, they send out an electrical signal (called an action potential) with the information "hey—we are smelling this specific chemical!" This signal gets passed along through nerves going into the olfactory bulb, up the olfactory peduncle, and into the forebrain. Then, the water flows back out of the rosette through the excurrent chamber which leads to the excurrent nostril and back out to the environment (Figure 3.2). Unlike humans, sharks do not sniff, instead all of this movement happens as the shark swims through the water (Figures 3.4 and 3.5).

Just as sharks come in a variety of shapes, sizes, habitats, and *trophic levels* (their place in the food web), their noses are also diverse! Some sharks have excurrent nostrils separate from their mouths, while others have it closer to or even directly connected to their mouths. Sharks have differently shaped olfactory rosettes too. Hammerheads (Sphyrnids) have long tubular noses while sharks with the typical pointy heads—including

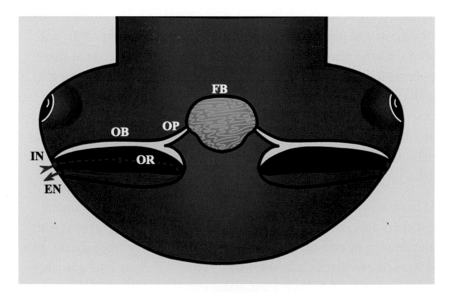

Figure 3.4 Olfactory system of a bonnethead shark (Sphyrna tiburo). *Blue arrows signify direction of water flow. EN – excurrent nostril, FB – forebrain, IN – incurrent nostril, OB olfactory bulb, OP – olfactory peduncle, OR – olfactory rosette.*

A. **B.**

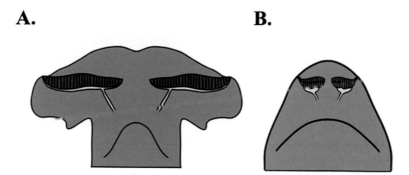

Figure 3.5 *Olfactory morphologies of the (A) small eye hammerhead (*Sphyrna tudes*) and the (B) lemon shark (*Negaprion brevirostris*).*

the mackerel (Lamnids), dogfish (Squalids), and requiem sharks (Carcharhinids) have stubbier, bulbous shaped noses (Figure 3.3).

While most sharks cover their olfactory rosettes, some of their close relatives (i.e. guitarfish) have rosettes that are directly exposed to the environment (Cox, 2013). This diversity continues in all aspects of olfactory morphology with the olfactory bulb size, number of lamellae, and rosette size (Schluessel et al., 2008; Yopak et al., 2015). Sharks also differ in how much of their lamellae are covered in sensory epithelium (the tissue that houses the olfactory receptor neurons).

Now, the question is, what do these differences mean? Several studies have looked at phylogeny (evolutionary relationships, i.e. who is related to who, as discussed in Chapter 2) and ecology (where it lives, what it eats, etc., also discussed in Chapter 2) and found that olfactory morphology is tied much more to the latter. For example, sharks with the largest olfactory bulbs live out in the wide open ocean environments while those associated with coral reefs have the smallest (Yopak et al., 2015). Sharks and their relatives that live at a variety of water depths (the bentho-pelagic elasmobranchs) also have more lamellae, more sensory epithelium coverage, and bigger rosettes than species living mostly on the ocean floor (benthic species). Additionally, these differences also correlate to diet. Elasmobranchs that eat food like clams and starfish have larger bulbs than those that eat food like crab (Schluessel et al., 2008).

Olfactory structure was previously believed to correlate with ecology. This led scientists to believe that sharks with bigger noses were better at smelling. However, work by Drs. Tricia Meredith and Stephen Kajiura have found that this isn't the case. Meredith and Kajiura (2010) found that in five species of sharks, there was no significant correlation between nose size and the ability to smell. However, the shape of the nose does have an impact on where the sensory structures lie. The shape of the nose impacts how water flows through it, and there is a correlation between water flow and nose sensitivity. The parts of the nose that receive less water flow have less sensory area than the parts that receive higher water flow. While this has been shown morphologically, further testing should look at if different parts of the nose are actually differently sensitive or whether it just looks that way (Simonitis and Marshall, 2022)

THE ENCOUNTER PT. 2: ELECTRIC BOOGALOO—SHORT RANGE SENSORY MORPHOLOGY AND PHYSIOLOGY

Once a shark has utilized its long range senses of magnetoreception, hearing, and olfaction to detect its food source, it has a second group of sensory superpowers at its disposal. In the aquatic environment, vision is a short range sense due to the difficulties of seeing in water. Light attenuates quickly in water, meaning that as it travels deeper in the water column, it is scattered and absorbed. For example, red light is quickly absorbed while blue light penetrates deeper. Additionally, water conditions are variable and high water movement could lead to a kick up of sediment, making it too murky of an environment to utilize vision well. Luckily, there are two other short range senses that sharks can employ either in addition to vision or to compensate for lack of vision in unfavorable water conditions

MECHANOSENSATION

In humans, mechanosensation (the sense of touch) is largely experienced through the use of hands and fingers. In order to feel the texture of a fabric, grip and squeeze an avocado to feel its firmness, or to press and manipulate Play Doh, receptors located under the skin of fingers carry signals to the brain. While human touch receptors are mostly located in the skin of appendages, shark skin looks rather different from human skin.

Shark skin is unique—they are covered in scales called dermal denticles which are exclusively found in elasmobranchs (sharks, skates, and rays). They cover the entire body, layering over each other much like the hair does on a cat or dog. These scales are believed to have evolved from teeth as they share morphological similarities. Structurally, denticles have a crown, a neck, and are attached to the body via a base which is embedded into the outer layer of skin. Like teeth, denticles also have layers of dentine, enamel, and an inner cavity (Oeffner and Lauder, 2012).

Generally, the purposes of dermal denticles are divided into four categories: protection from abrasion (cuts and scratches), defense from external objects (living or nonliving), bioluminescence (the ability to produce light), and drag reduction when moving through water (Reif, 1978). Denticles are hard, and thus protect and defend the shark's skin (and the sensory organs located in the skin) from abrasions and injuries. In 2020, it was discovered that whale sharks even have dermal denticles on their eyes likely for the purpose of protection (Tomita et al., 2020). They are the only known species to have this adaptation.

The physical properties of denticles such as size, shape, and orientation influence how the body of the shark interacts with the surrounding water. Their size and shape vary across species and across one individual shark's body (Sullivan & Regan, 2011). The denticles can be rounded or pointed and the surface can have ridges or be flat. Denticle structure allows water to move smoothly over the shark's body, reducing drag—the forces experienced on the body in the opposite direction that the shark is swimming. Drag reduction allows for less energy expenditure while swimming (Domel et al., 2018).

Among and underneath the dermal-denticle-covered skin lies the flow-sensing (mechanosensory) system in sharks: the lateral line. The lateral line is comprised of sensory organs called neuromasts that detect water flows (De la Cruz-Torres et al., 2018). This system is found in both bony fishes and cartilaginous fishes. In elasmobranchs, neuromasts are found in specific patterns on the head, and usually make up a single line that goes down the midline of the trunk to the end of the tail (Tester & Kendall, 1969). Neuromasts contain sensory hair cells, similar to those discussed in the hearing section. The cilia on the surface of the hair cells project into a gelatinous cap, called the cupula (Figure 3.6). When water flows over the cupula, it moves with the water, pushing the bundles of the hair cells underneath it. This causes an action potential to be sent to the brain of the shark via the nerves at the base of the hair cells. These signals from the lateral line system help fish detect water movement and changes in the surrounding water pressure in their environment (Mogdans, 2019). This system plays a key role in prey detection, but may also be used by groups of sharks (aka "shivers") to avoid bumping into each other.

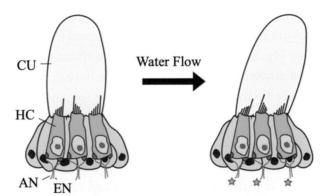

Figure 3.6 *A neuromast sensory receptor organ of the lateral line system made up of hair cells (blue, HC). Each hair cell is innervated by afferent (AN) and efferent (EN) nerves. The hair cells have long kinocilia and shorter stereovilli that are projecting into the gelatinous cupula (CU). When the neuromast is deflected by a water flow, the cupula is pushed over which pushes the cilia of the hair cells over. This activates the hair cells and sends a signal (represented by stars) to the brain of the fish via the afferent nerves. Modified from Gardiner & Atema, 2014.*

There are two types of neuromasts found on sharks: superficial and canal neuromasts (Figure 3.7). Superficial neuromasts are found on the skin underneath the ridges of dermal denticles. They sense the direction and how fast water flows past the body. The second type of neuromast, canal neuromasts, sit in hollow, fluid filled canals under the surface of the skin that are located in specific locations across the head, trunk, and all the way down to the tip of the tail. Canal neuromasts are connected to the external environment by pores and can detect changes in the pressure gradient. A pressure gradient is the difference in pressure between two pores caused by a change in the water flow around a shark or ray's body (Zhai et al., 2021). Denticles surround pores and can grow around the pore. Sometimes the denticle will cover the pore leaving a small amount of space between the crown of the denticle itself and the opening of the pore (Gardiner & Atema, 2014; Peach & Marshall, 2009).

Elasmobranchs tend to have canals on both the dorsal and ventral sides of their head. However some species, such as the torpedo rays (of the Order Torpediniformes), only have canals in their dorsal region (De la Cruz-Torres et al., 2018). Some species of ray have non-pored canals on the ventral side of their head, which have been suggested to function as touch receptors. These non-pored canals may be important for feeding

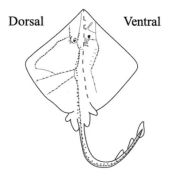

Dorsal   Ventral

Figure 3.7 *The lateral line system of the clearnose skate,* Raja eglanteria. *The dorsal side on the left contains lateral line canals (gray lines), canal neuromasts (red lines), and superficial neuromasts (blue dots) from the head down to the length of the tail. The ventral side on the right shows the canals (gray lines) and the location of their respective neuromasts (red dots). The dashed line down the middle annotates the division between the dorsal and ventral sides in this illustration. Adapted from Maruska 2001.*

along the bottom of the sea (Maruska & Tricas, 1998). Meanwhile, superficial neuromasts are not limited to the location of canals, and can be located all over the head. Since neuromasts have directional sensitivity, they can give sharks information about the direction that the water flow signal is coming from.

The lateral line system plays an important role in prey detection. Different types of prey produce specific types of water movement. Fish are able to tell whether water movement is being produced by an animal or a water current (Mogdans, 2019). The lateral line system can also be used in tandem with the olfactory system to detect the location of odor sources, since smells are carried by flowing water. The movement of water detected by the lateral line system gives the fish an idea of which direction the smell is coming from (Gardiner & Atema, 2014). Without water flowing over the lateral line, a predator would not be able to know which direction the smells are coming from.

There are, of course, many things we still do not know about the lateral line system. While most of the information that exists regarding lateral line function is known for bony fishes, very few studies have examined lateral line function in sharks and rays. In bony fishes, the lateral line is known to play a role in predator avoidance, social communication (particularly for mating), schooling with other fish, and water flow orientation around objects (like a salmon holding position behind a rock in a river). None of these functions have been experimentally demonstrated in elasmobranchs, leaving this field wide open for future research. Scientists also have yet to find out how the changing water conditions may affect a shark's ability to sense water flow. There is a possibility that the lateral line system changes, either through short-term or long-term changes over several generations in response to a changing environment (Mogdans, 2019). Since there are similar hair cells in both ears and the lateral line system, neuromasts share some of their overlapping sensitivity ranges with the inner ear of sharks. Just like we see in hearing, human-produced noise may also have negative effects on the lateral line function as well (Chaupis et al., 2019). All of these areas of potential future research will continue to contribute to understanding the importance of the lateral line system to elasmobranchs and shed light on how sharks and rays cope with the impacts of climate change.

ELECTRORECEPTION

Much of the hype around shark sensory biology stems from their "sixth sense," or their ability to sense weak electric fields through electroreception. They do this through the use of *ampullae of Lorenzini*, a type of ampullary organ (electrosensing organ) that can detect very weak electric fields generated by living organisms (Newton et al., 2019). Electroreceptive organs have evolved over 500 million years in aquatic animals and have changed, been lost, and re-evolved independently many times (Kempster et al., 2012). Although other types of ampullary organs are found in other kinds of fish, ampullae of Lorenzini refer specifically to the ampullary receptors found in sharks and their cartilaginous relatives. Ampullae of Lorenzini were first described by Stenonis in 1664 and their namesake Lorenzini in 1678 (England & Robert, 2021; Newton et al., 2019). This electrosensory system plays an important role during feeding events. Sharks, skates, and rays use their ampullae of Lorenzini to detect, orient towards, and capture prey.

The pores of the ampullae of Lorenzini can be seen externally, mostly on the head, of elasmobranchs (Figure 3.8). Each of these pores leads to a

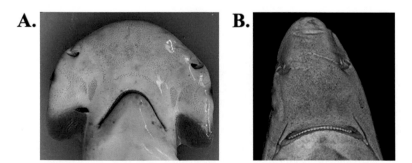

Figure 3.8 *(A) Pores of ampullae of Lorenzini on the underside of a bonnethead shark (Sphyrna tiburo) and (B) a Pacific spiny dogfish (Squalus suckleyi) microCT scanned and stained with 5% phosphomolybdic acid. Scanned and reconstructed by Dr Lauren Simonitis (l-aps:LES:SS-1; MorphoSource ARK ark:/87602/m4/430669).*

canal filled with a jelly substance that connects to the ampulla. The jelly in the canal is extremely conductive, meaning it has an amazing capacity to pass electrical signals. This ampulla is the part of the structure where the electric signals are detected by the receptors. The ampulla is filled with sensory cells similar to the hair cells previously mentioned in both the mechanosensation and hearing sections of this chapter (Figure 3.9). However, unlike hair cells in the inner ear and lateral line, they only have one long cilium rather than many cilia. When an electric stimulus, such as a muscle contraction in a fish, is produced, these signals travel through seawater. When a shark or ray comes into contact with this signal, it is transferred through the pore, travels through that conductive jelly in the canal, and stimulates the ampulla's sensory cells. The sensory cells detect the electrical currents and send an action potential (via nerves) to the brain. (Kempster et al., 2012; Kozak & López-Schier, 2017; Newton et al., 2019).

Ampullae of Lorenzini even allow sharks to locate potential prey that might be obscured from their other sensory systems, such as prey out of their line of sight (Kajiura & Holland, 2002). This is especially true

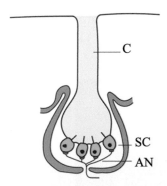

Figure 3.9 *An Ampullae of Lorenzini electroreceptive organ of an elasmobranch. The canal (C) sits below the surface of the skin and is filled with jelly that conducts electric signals. The sensory cells (SC) have single cilia that when stimulated by an electrical signal activates the sensory cells that send a signal to the brain via the afferent nerve (AN). This is a simple ampullary organ, but some elasmobranchs have multiple pores for each ampule, several ampules for each organ, and the length of the canal can vary greatly. Adapted from Alves-Gomes 2001.*

for hammerhead sharks, rays, and skates, who can detect prey buried underneath them, even though their eyes are on top of their head (rays and skates) or on the far ends of their long heads (hammerheads). This amazing system is what allows sharks to hone in on a biopotential signal, like the beating of a flounder's heart buried under the sand or the muscle contractions of a panicking or injured fish. Because their ampullae of Lorenzini are most useful while hunting at close range, the highest density of pores is generally found near the mouth (Newton et al., 2019).

Ampullae of Lorenzini are found in all elasmobranch species, though the complexity and distribution of the electroreceptors vary. In sharks, they're mostly on the head but they extend onto the body and wings of rays and skates (Jordan, 2008; Newton et al., 2019). The electroreceptive organs in both sharks, skates, and rays are arranged in different patterns or clusters, and can also vary in size between species. Similar to other sensory systems, the structure and function of the electrosensory organs is more related to the elasmobranch's ecology (where it lives, what it eats, who eats it, etc.) than who it's related to evolutionarily. Benthic-feeding sharks and skates who feed close to the ocean floor have a higher number and density of pores on the underside of the head, while pelagic sharks feeding out in the open ocean have fewer pores, lower densities, and more equal distributions between the top and bottom side of their heads (Jordan,2008). The pores on hammerhead sharks (Family Sphyrnidae) are arranged in clusters spread across the large extensions of their cephalofoil (their weird, elongated heads). Because they have such a wide surface for picking up signals across their cephalofoil, they can pinpoint the direction the signal is coming from. For example, if they feel a stronger voltage signal on one side of their head, they know that the source is closer to the end with the stronger voltage. This is also believed to give them a better electrical field detection ability and allow them to search for prey over a larger area. It has been hypothesized that enhanced electroreception might have driven the evolution of the hammerhead head shape and that this head shape provides greater electrosensory capabilities. A study conducted by Kaijura and Holland (2002) was the first to test this hypothesis, and found that hammerhead sharks sample a greater overall area due to their greater head width, which increases their probability of a prey encounter.

Sharks are born with a finite amount of electrosensory pores. As they grow, they do not gain more or redistribute them. This means that the resolution (the number of receptors in a certain amount of space) of the electrosensory system decreases as a shark grows (Kajiura, 2001; Newton et al., 2019). As the individual grows, the exterior pore also grows further away from the ampulla and its hair cells. However, lengthening of the canal actually results in an increase in receptor sensitivity (Kajiura, 2001; Newton et al., 2019). Furthermore, a study conducted by Kajiura (2003) documented how well neonatal (newborn) bonnethead sharks respond to prey-generated electric stimuli. These baby sharks exhibited responses only 24 hours after birth! This study was the first to examine electroreception in newborn sharks and demonstrates that they possess a well-developed electrosensory system enabling them to detect prey soon after birth (Kajiura, 2003). A question that still remains is whether this high degree of sensitivity increases throughout all stages of life as it does for other elasmobranch species (Kajiura, 2003).

While we understand some functional aspects of shark electroreception, a number of questions remain. Scientists are still studying electrosensory anatomy, physiology, and behavior and how these relate to species habitat, diet, and overall morphology. Studies on electroreception in elasmobranchs have focused on species that are easily kept in captivity and used often in the lab, which has resulted in an overrepresentation of just a few species and gaps in knowledge about other species (Newton et al., 2019). However, in the future, researchers can take what is known about these well-studied species and make reasonable assumptions about underrepresented species based on what they know from the foundational work. While past research has created a solid foundation for understanding electroreception in elasmobranchs, there is still more to learn and many questions that remain to be tested by future sensory biologists.

THE ATTACK—MUSCLE MORPHOLOGY AND PHYSIOLOGY

Now, at this point in a predation event, the shark has used both its long range and short range senses to pinpoint its prey item. Now begins the next phase of the predation event—the attack. This part of the predation event really highlights the shark's muscles allowing it to move in for the kill. Elasmobranchs, like bony fishes and even mammals (including humans), have several types of muscles. Muscle is an important body tissue present within most organ systems, but skeletal muscle is the largest in shear mass of musculature in a fish's body. In some fishes, skeletal muscle comprises almost 60% of their body mass (Listrat et al., 2015). Different movements require activation of different muscle groups—whether a fish is cruising through the water at low speeds, resting on the seafloor, accelerating towards prey, or capturing prey! Muscle physiology and motor activity of elasmobranchs are less well-studied than in bony fishes, but there is still enough research to get an idea of what is occurring and provokes the need to study more species.

First, here is a breakdown of the different types of skeletal muscle present in many elasmobranchs. Just like humans, elasmobranchs have red muscle used for aerobic ("with oxygen") *metabolism* and movement and white muscle used for anaerobic ("without oxygen") metabolism and burst movement. Red and white muscles are named after their color, which is based on the amount of *myoglobin* present in the muscle. Myoglobin is a really important molecule which stores and transports oxygen within the muscle. Red muscle has a much higher myoglobin content than white muscle, which makes sense given that it is important for aerobic (oxygen-dependent) movement. Red muscle also contains large amounts of mitochondria, which produce oxygen and are important for energy production. White muscle does not have myoglobin, but does contain mitochondria, though not at nearly the same density as red muscle. White muscle also has fewer blood vessels and less blood flow leading to less oxygen being brought to the tissue compared to red muscle. Pink muscle is considered intermediate between red and white muscle and is more often present in open ocean elasmobranchs who have to travel long distances. Where these muscle types occur in the body and the proportion of skeletal muscle they comprise is important in the context of an elasmobranch's habitat and lifestyle. Most skeletal muscle of fishes is white muscle and a very small amount of red muscle because red muscle requires more energy to maintain. Many fish don't actually need that much red muscle for long, sustained movement.

Internal temperature is an important component of the energetic demand for elasmobranchs and relates to the activity of the species and the muscle type used. While most elasmobranchs are ectothermic, meaning their internal body temperature is the same as the surrounding environment, some employ *regional endothermy*—they can maintain higher temperatures within their body than the surrounding environment in specific body areas, such as internalized red muscle (Figure 3.10). One of the most popular examples of this is the shortfin mako shark, *Isurus oxyrinchus*, known for traveling across extreme temperature changes in the ocean. They are able to do this because they capture heat produced in a metabolic process called vascular countercurrent exchange (Harding et al., 2021). Vascular countercurrent exchange works by having warm blood, heated by the body's core, pass near cold blood, which has lost heat from supplying oxygen to the extremities away from the body's core. Normally, blood travels from the warm core of the body to the extremities. As it moves away from the core, it loses heat and becomes cooler. Once the blood comes back from delivering oxygen to the body's extremities, it is colder and closer to the external temperature, lowering the body temperature of the muscles around it. In regional endothermy, this colder returning blood passes close to the warm blood, allowing for heat transfer (Figure 3.11). This vascular countercurrent exchange directly benefits red muscle with thermal stability for sustained swimming (Syme & Shadwick, 2011). Other species who can regulate their internal temperature include the common thresher shark (*Alopias vulpinis*) and the salmon shark (*Lamna ditropis*; Bernal et al., 2003; Bernal et al., 2012; Donley et al., 2012;

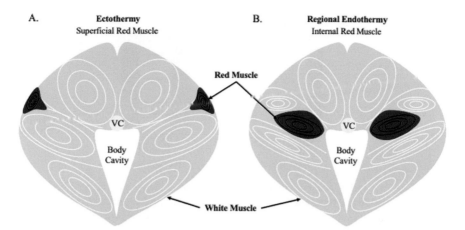

Figure 3.10 *Ectothermy and Regional Endothermy in Skeletal Muscle of Sharks. (A) Representative diagram of ectothermy in sharks, characterized by red skeletal muscle located laterally, just beneath the skin and the majority of locomotory muscle is white muscle. This represents the general muscle profile of most shark species studied to date. (B) Representative diagram of regional endothermy in sharks, characterized by red skeletal muscle located deep in the body towards the body cavity. A majority of the skeletal muscle in regionally endothermic sharks is also white skeletal muscle and this general profile represents species such as the shortfin mako (*Isurus oxyrinchus*), common thresher (*Alopias vulpinus*), and salmon sharks (*Lamna ditropis*). VC= Vertebral Column.*

Patterson et al., 2011). Regional endothermy is best for those large fish species that need to migrate long distances for food and habitat resources because it allows for faster speeds and greater muscle power output in the midst of variable temperatures and environments (Shadwick & Goldbogen, 2012; Syme & Shadwick, 2011).

Think back to that large, endothermic, pelagic shortfin mako lunging forward to grasp its prey and launch itself out of the water. To achieve this, they use bursts of speed up to 10 m/s, thrusting them out of the water. This is an anaerobic movement that would recruit white skeletal muscle for power (Bernal et al., 2003). However, due to the regional endothermy of the mako shark, the combined aerobic and anaerobic capacities of their white muscle are greater than in species without regional endothermy (Dickson et al., 1993). On the opposite end of the spectrum, the Greenland shark (*Somniosus microcephalus*) is known for its slow movements and "burst" speeds of .74 m/s. Despite their

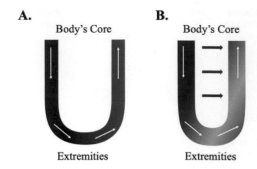

A. Body's Core
B. Body's Core

Extremities
Extremities

Figure 3.11 *In a system without vascular countercurrent exchange (A), warm blood leaves the body's core, loses heat at the extremities, and returns to the core cold. However, in a system with vascular countercurrent exchange (B) the cold blood returning from the extremities passes closer to the warm blood leaving the body's core, allowing for heat transfer. This warms the blood returning to the body. Blue = cooler temperatures, red = warmer temperatures.*

slow moving lifestyle, they somehow can capture fish, invertebrates, and large land mammals such as Arctic seals (Watanabe et al., 2012). The musculature profile of this species has still not been discussed in the literature! Scientists predict that their white skeletal musculature is limited not only by their sheer body mass but also by the frigid temperatures where Greenland sharks are known to reside (Watanabe et al., 2012).

Skeletal muscle isn't the only musculature recruited during prey capture, it is just the beginning of the sequence of attack. We know that the cranial, jaw, and hyoid muscles are all involved in the feeding behavior of many elasmobranchs from sharks to batoids. Studies of feeding behavior on mackerel (Lamnids), dogfish (Squalids), and requiem sharks (Carcharhinids) generally show a conserved feeding kinematic sequence, which is a pattern of movement or energy transfer through structures used for feeding (Dean et al., 2005; Frazzetta & Prange, 1987; Motta et al., 1997; Wilga & Motta, 1998a). In the spiny dogfish (*Squalus acanthias*), some jaw retractor muscles are active during the expansion or the recovery of the jaw depending on the feeding behavior (capture, manipulation, and transport; Wilga & Motta, 1998a). The Atlantic guitarfish (*Rhinobatos lentiginosus*) shows similar muscle function and locomotor patterns in the head muscle as spiny dogfish (Wilga & Motta, 1998b).

Some elasmobranchs even use their fins for feeding. Rays recruit both red and white muscles for the undulating (up and down) movements of their flappy pectoral fins, which is called "rajiform locomotion." Many rays use their pectoral fins for feeding and locomotion because of the flexible nature of their muscles and skeleton. Eagle rays specifically have extensions on their pectoral fins called "cephalic lobes," which are used to help direct prey into their mouths in open ocean environments.

The research focused on muscle morphology and physiology of a range of elasmobranchs is ongoing, especially with regard to feeding and prey encounters. Understanding muscle activity during various movements and behaviors helps scientists better understand how much energy individuals are using to acquire resources. More research is needed on musculature composition, physiology, and motor activity in a broader range of elasmobranchs.

THE CAPTURE—JAW MORPHOLOGY AND PHYSIOLOGY

Now that a shark has locked on to its prey with its senses and moved in for the attack with its muscles, it's time to capture the prey with their mouths. Like humans, sharks have jaws and teeth that assist with eating. There are multiple ways that sharks feed such as suction, cutting, gouging, crushing, and filter feeding. Depending on the way a shark eats, their upper and lower jaw positions and surrounding anatomy can vary. The first sharks to appear on earth had a jaw set more forward on their face with the nose and mouth positioned directly above and below each other (Wilga, 2005). As sharks evolved, their upper and lower jaws moved to be more ventral (towards the underside), which set the nose as the first sense or point of contact on the face before the mouth. Why might this have occurred? This is likely a preferred set up since the nose can sample the environment before the mouth (Moss, 1972). This also gives space for ampullae of Lorenzini and the lateral line—both crucial mechanisms for sharks in searching for prey, which were discussed earlier in the chapter.

In modern sharks there are three types of jaw suspensions: amphistylyic, orbitostylic, and hyostylic (Figure 3.12). These three jaw morphologies are distinct from each other in that they have different cartilaginous structures and attachments to the skull. Amphistylyic jaw suspension is seen in the most primitive (oldest) order of sharks alive today, hexanchiformes or the cow sharks as well as in shark fossils. Hyostylic and obitosylic jaw suspension differ from amphistylic because the lower jaw can project at varying angles from the skull (Abel & Grubbs, 2020).

These three jaw suspensions directly correlate with how a shark eats and what it eats. Hyostylic and orbitostylic jaw suspension are ideal for gouging bites out of their prey. (Wilga, 2005). Requiem sharks (e.g. tiger sharks (*Galeocerdo cuvier*) and bull sharks (*Carcharhinus leucas*)) and dogfish are known for this gouging. These sharks are also known to shark their heads back and forth as they bite, exhibiting a cutting behavior. Suction feeders, such as whale sharks (*Rhincodon typus*), have specialized musculature within the mouth that allows them to suck water and prey items in. Suction feeding sharks have shorter jaws and a smaller gape (how wide the mouth can open) than non-suction feeding sharks (Moss, 1972). Nurse sharks (*Ginglymostoma cirratum*) are another great example of a suction feeder that has specific muscles and cartilage within their face that allow them to suck in and subsequently crush their prey.

Jaw suspension does not act alone in allowing sharks to eat their prey. Just like in humans, teeth help break down food items into smaller sections that can be swallowed. Tooth shape varies broadly depending on the shark and is indicative of how that shark eats. A shark who gouges or cuts will have a different tooth shape than a shark who crushes its prey

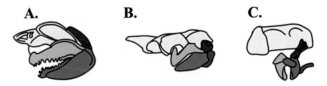

A. **B.** **C.**

Figure 3.12 *Examples of shark jaw suspension, including: (A) Amphistylic, as seen in Lamnidae, (B) Orbitostylic, as seen in Squalidae, and (C) Euhyostylic, as seen in Rajidae. Modified from Wilga, 2005.*

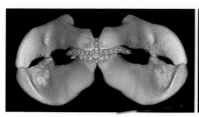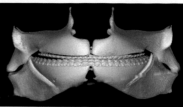

Figure 3.13 MicroCT scanned jaws of a Japanese bullhead shark *(*Heterodontus japonicus, *left) and pacific spiny dogfish (*Squalus suckleyii, *right). Scanned by Dr Matthew Kolmann (sio:marine vertebrates:70379; MorphoSource ARK ark:/87602/ m4/430664) and Kayla Hall (l-aps:KCH:SS-1; MorphoSource ARK ark:/87602/ m4/430658). Reconstructed by Dr Lauren Simonitis.*

item. Sharks do not chew their food and often must shake their head back and forth to separate a bite from the prey. This often results in the breakage or loss of teeth. As a result, sharks employ tooth replacement— when one tooth is lost, another replaces it. A single shark can go through thousands of teeth a year. The number of rows of replacement teeth as well as the replacement rate vary broadly, ranging from a few days to a few weeks and from a few rows to tens of rows (Figure 3.13).

There are some weirdos in the bunch that are worth highlighting. Cookiecutter sharks (*Isistius brasiliensis*) replace their entire bottom row of teeth at once, swallowing it whole, a practice scientists believe to be a way of recycling nutrients in its nutrient-poor habitat (Strasburg, 1963). Helicoprion, a prehistoric and distant relative to sharks, replaced teeth like a conveyor belt with each row of teeth folding back into the jaw to create the spiral jaw fossils we continue to discover today (Tapanila et al., 2013). Imagine if instead of losing a baby tooth and then having only one replacement adult tooth, humans could lose and replace as many teeth as they wanted!

THE INGESTION—DIGESTIVE MORPHOLOGY AND PHYSIOLOGY

Finally, the shark has reached the final phase of the predation event. After a successful prey detection, attack, and capture, the shark is ready to digest its hard-earned meal. Sharks occupy a variety of different places in the food web and eat everything from zooplankton, crustaceans, echinoderms, cephalopods, bony fish, marine mammals, even other sharks. Tiger sharks (*Galeocerdo cuvier*) have even been documented eating migratory birds who have fallen from the sky (Drymon et al., 2019). One species of shark, the bonnethead shark (*Sphyrna tiburo*), has actually been shown to digest seagrass! Bonnethead sharks swim along seagrass beds, chomping up the critters within. Because of this, researchers have found large amounts of seagrass in their stomach contents. Recently, work by Drs Samantha Leigh, Yannis Papastamatiou, and Donovan German has shown that they are actually absorbing nutrients from seagrass, making bonnetheads the first known omnivorous sharks (Leigh et al., 2018).

Elasmobranchs have what is known as a complete digestive system meaning that the entry point for the digestive tract is separate from the exit point. First, the prey enters the oral cavity (mouth) through different ingestion methods like filter, ambush, suction, or gape feeding. Benthic, or bottom-dwelling, predators are commonly seen utilizing suction feeding for prey ingestion. Filter feeding sharks, like whale sharks (*Rhincodon typus*) and basking sharks (*Cetorhinus maximus*), target plankton—a diverse group of free-drifting microscopic organisms. By suctioning in the water and passing it through a filtering apparatus, the plankton are able to be separated from the water column and ingested. Other sharks employ ambush feeding. Utilizing stealth and rapid bursts of speed, these ambushing sharks are able to capture their prey. A review of these different feeding types can be found in the above section about jaw morphology. Once in the buccal cavity, the prey is passed through into the predator's esophagus. The esophagus is composed of striated muscles that secretes mucus to aid in the transfer of the prey to the stomach (Leigh et al., 2017).

Throughout the process of digestion, the predator acquires biological and organic molecules through metabolism. By metabolizing the food that they eat, important components such as carbohydrates, proteins, and lipids are gathered and converted into usable energy. Elasmobranchs are ectothermic, so their internal processes are often affected by the ambient conditions of their surroundings. As the temperature increases, the rate of their metabolism will increase and vice versa. Research by Lorena Silva-Garay and Dr Christopher Lowe demonstrated that stingrays, especially juveniles, were very metabolically sensitive to changes in temperature with their metabolic rate decreasing as temperature increases. This poses the question of how these species will ultimately be affected by the increase in water temperature due to climate change (Silva-Garay & Lowe, 2021)

Following the stomach are the anterior and posterior intestines. The inner lining of the intestines can be found in an array of folds and a spiral valve that maximizes the surface area for the newly obtained nutrients. The presence of the spiral valve is unique to elasmobranchs. The spiral valve is actually hypothesized to be an example of a Tesla valve—meaning it allows fluid to flow in one direction passively without any moving parts (Leigh et al., 2021)! Much like humans, the gastrointestinal tracts within sharks, rays, and skates include a microbiome, a collection of microorganisms, that aids in the digestive capabilities of vertebrates (Leigh et al., 2021). As previously stated, bonnethead sharks consume seagrass as a part of their diet. A primary component within plant cells is cellulose, an organic compound that is indigestible without the presence of the enzyme cellulase. While this enzyme is not produced within animal cells, the microbes within the system of bonnethead sharks allow them to be the only known shark capable of digesting plant material (Leigh et al., 2021). Once the maximum amount of nutrients are obtained, what is left is excreted through the animal's cloaca and out of the digestive tract. Sharks also have a special structure called a "rectal gland," which helps them excrete excess salt from their bodies. Recent work by Drs Jinae Roa and Martin Tresguerres found that one of the same acid-base sensors and regulators that is in the mammalian kidney is found in the rectal gland of the leopard shark (*Triakis semifasciata*; Roa & Tresguerres, 2017) (Figure 3.14).

Understanding the process of digestion enhances the knowledge of what it takes for these species to survive when they don't know when they

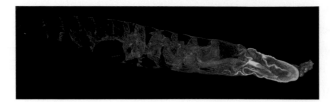

Figure 3.14 Cross section through a microCT scanned spiral valve of a gulper shark (Centrophorus granulosus). Scanned by Dr Samantha Leigh (l-aps:05:1; MorphoSource ARK ark:/87602/m4/M168177). Reconstructed by Dr Lauren Simonitis.

will get their next meal. As has been the theme of this chapter, scientists know more about digestive morphology and physiology in bony fish than elasmobranchs. Additionally, there are many different sharks living in a variety of extremely different environments and eating lots of different prey items. For many of these species, their digestive system is severely understudied leaving the door open for future digestive morphologists and physiologists to dive right in and get to work bridging these knowledge gaps.

CONCLUSIONS

This is just a brief overview of the current knowledge of shark anatomy and physiology. This chapter has detailed the basics of how sharks look and function during a predation event. Research has been presented over their sensory capabilities used for detecting prey, their muscles powering them through the water as they approach, their specialized jaws for catching their food, and their digestive system for gaining nutrients and energy. These four areas of anatomy and physiology are just a fraction of what makes sharks successful in their environment.

This chapter has also highlighted what questions still need to be answered. Especially when compared to bony fish, our knowledge of shark morphology and physiology is lacking. Sensory biologists are still filled to the brim with questions: can sharks detect the pressure component of sound? Do sharks with differently shaped noses have different distributions of smell-sensing areas? How does the lateral line help sharks form and coordinate while swimming in shivers? How does a shark's ability to sense electricity change over their lifetime? Muscle physiologists are still puzzled by how the slow moving Greenland shark is able to capture such fast prey. Morphologists and biomechanists are still investigating how the jaws and teeth interface in sharks with different feeding strategies. Digestive physiologists are still discovering new ways that sharks gain nutrients from their food—both animal and plant material sometimes!

Additionally out of the more than 500 species of sharks, only a select few are represented here. Because anatomy and physiology vary along with ecology, it is important to conduct comparative studies on sharks with differing habitats, diets, behaviors, etc. Just because one of these questions was answered for a select few shark species does not mean it is applicable to all of them. The field is wide open for upcoming shark scientists who want to study the form and function of sharks and their relatives.

REFERENCES

Abel, D. C., & Grubbs, R. D. (2020). *Shark biology and conservation: Essentials for educators, students, and enthusiasts.* Baltimore, MD: Johns Hopkins University Press.

Alves-Gomes, J. A. (2001). The evolution of electroreception and bioelectrogenesis in teleost fish: A phylogenetic perspective. *Journal of Fish Biology, 58*(6), 1489–1511. https://doi.org/10.1111/j.1095-8649.2001. tb02307.x.

Banner, A. (1968). Attraction of young lemon sharks, Negaprion brevirostris, by sound. *Copeia, 1968*(4), 871–872.

Banner, A. (1972). Use of sound in predation by young lemon sharks, Negaprion brevirostris (Poey). *Bulletin of Marine Science, 22*(2), 251–283.

Barber, V. C. et al. (1985). Quantitative analyses of sex and size differences in the macula neglecta and ramus neglectus in the inner ear of the skate, Raja ocellata. *Cell and Tissue Research, 241*(3), 597–605.

Bernal, D. et al. (2012). Energetics, metabolism, and endothermy in sharks and rays. In J. C. Carrier, J. A. Musick, & M. R. Heithaus (Eds.), *Biology of sharks and their relatives* (Vol. 211, p. 237). Boca Raton, FL: CRC Press.

Bernal, D., Sepulveda, C., Mathieu-Costello, O., & Graham, J. B. (2003). Comparative studies of high performance swimming in sharks I. Red muscle morphometrics, vascularization and ultrastructure. *Journal of Experimental Biology, 206*(16), 2831–2843.

Chapuis, L., Collin, S. P., Yopak, K. E., McCauley, R. D., Kempster, R. M., Ryan, L. A., … Hart, N. S. (2019). The effect of underwater sounds on shark behaviour. *Scientific Reports, 9*(1), 1–11.

Corwin, J. T. (1977). Morphology of the macula neglecta in sharks of the genus Carcharhinus. *Journal of Morphology, 152*(3), 341–361.

Corwin, J. T. (1981). Audition in elasmobranchs. In W. N. Tavolga, A. N. Popper, & R. R. Fay (Eds.), *Hearing and sound communication in fishes* (pp. 81–105). Berlin: Springer.

Cox, J. P. L. (2013). Ciliary function in the olfactory organs of sharks and rays. *Fish and Fisheries, 14*(3), 364–390. https://doi.org/10.1111/j.1467-2979.2012.00476.x.

De la Cruz-Torres, J., González-Acosta, A. F., & Martínez-Pérez, J. A. (2018). Descripción y comparación de la línea lateral de tres especies de rayas eléctricas del género Narcine (Torpediniformes: Narcinidae). *Revista de Biología Tropical, 66*(2 June), 586–592. https://doi.org/10.15517/rbt.v66i2.33382.

Dean, M. N., Wilga, C. D., & Summers, A. P. (2005). Eating without hands or tongue: Specialization, elaboration and the evolution of prey processing mechanisms in cartilaginous fishes. *Biology Letters, 1*(3), 357–361.

Dickson, K. A., Gregorio, M. O., Gruber, S. J., Loefler, K. L., Tran, M., & Terrell, C. (1993). Biochemical indices of aerobic and anaerobic capacity in muscle tissues of California elasmobranch fishes differing in typical activity level. *Marine Biology, 117*(2), 185–193.

Domel, A. G., Domel, G., Weaver, J. C., Saadat, M., Bertoldi, K., & Lauder, G. V. (2018). Hydrodynamic properties of biomimetic shark skin: Effect of denticle size and swimming speed. *Institute of Physics, 5*(August), 056014. https://doi.org/10.1088/1748-3190/aad418.

Donley, J. M., Sepulveda, C. A., Aalbers, S. A., McGillivray, D. G., Syme, D. A., & Bernal, D. (2012). Effects of temperature on power output and contraction kinetics in the locomotor muscle of the regionally endothermic common thresher shark (Alopias vulpinus). *Fish Physiology and Biochemistry, 38*(5), 1507–1519.

Drymon, J. M., Feldheim, K., Fournier, A. M. V., Seubert, E. A., Jefferson, A. E., Kroetz, A. M., & Powers, S. P. (2019). Tiger sharks eat songbirds: Scavenging a windfall of nutrients from the sky. *Ecology, 9*(9), e02728.

England, S. J., & Robert, D. (2021). The ecology of electricity and electroreception. *Biological Reviews, 97*(1), 383–413.

Fay, R. R., Kendall, J. I., Popper, A. N., & Tester, A. L. (1974). Vibration detection by the macula neglecta of sharks. *Comparative Biochemistry and Physiology: Part A, 47*(4), 1235–1240.

Fekete, D. M. (2003). Rocks that roll zebrafish. *Science, 302*(5643), 241–242.

Frazzetta, T. H., & Prange, C. D. (1987). Movements of cephalic components during feeding in some requiem sharks (Carcharhiniformes: Carcharhinidae). *Copeia, 1987*(4), 979–993. https://www.jstor.org/stable /1445562#metadata_info_tab_contents.

Gardiner, J. M. et al. (2012). Sensory physiology and behavior of elasmobranchs. *Biology of sharks and their relatives, 1*, 349–401.

Gardiner, J. M., & Atema, J. (2014). Flow sensing in sharks: Lateral line contributions to navigation and prey capture. In H. Bleckmann et al. (Eds.), *Flow sensing in air and water: Behavioral, neural and engineering principles of operation* (pp. 127–146). Berlin: Springer. https://doi.org/10.1007/978-3 -642-41446-6_5.

Harding, L., Jackson, A., Barnett, A., Donohue, I., Halsey, L., Huveneers, C., … Payne, N. (2021). Endothermy makes fishes faster but does not expand their thermal niche. *Functional Ecology, 35*(9), 1951–1959.

Jordan, L. K. (2008). Comparative morphology of stingray lateral line canal and electrosensory systems. *Journal of Morphology, 269*(11), 1325–1339.

Kajiura, S. M. (2001). Head morphology and electrosensory pore distribution of carcharhinid and sphyrnid sharks. *Environmental Biology of Fishes, 61*(2), 125–133.

Kajiura, S. M. (2003). Electroreception in neonatal bonnethead sharks, Sphyrna tiburo. *Marine Biology, 143*(3), 603–611.

Kajiura, S. M., & Holland, K. N. (2002). Electroreception in juvenile scalloped hammerhead and sandbar sharks. *Journal of Experimental Biology, 205*(23), 3609–3621.

Kempster, R. M., McCarthy, I. D., & Collin, S. P. (2012). Phylogenetic and ecological factors influencing the number and distribution of electroreceptors in elasmobranchs. *Journal of Fish Biology, 80*(5), 2055–2088.

Kozak, E. L., & López-Schier, H. (2017). Sensory systems: Electrifying news from the ocean. *Current Biology, 27*(24 December), R1327–R1329. https://doi.org/10.1016/j.cub.2017.11.015.

Leigh, S. C., Papastamatiou, Y., & German, D. P. (2017). The nutritional physiology of sharks. *Reviews in Fish Biology and Fisheries, 27*(3 September), 561–585. https://doi.org/10.1007/s11160-017-9481-2.

Leigh, S. C., Papastamatiou, Y. P., & German, D. P. (2018). Seagrass digestion by a notorious 'carnivore'. *Proceedings of the Royal Society of London: Series B, 285*(1886), 20181583.

Leigh, S. C., Summers Adam, P., Hoffmann Sarah, L., & German Donovan, P. (2021). Shark spiral intestines may operate as Tesla valves. *Proceedings Biological Sciences.* http://doi.org/10.1098/rspb.2021.1359

Leigh, S. C., Summers, A. P., Hoffmann, S. L., & German, D. P. *royalsocietypublishing.org (Atypon).* Retrieved from https://doi.org/10.1098/rspb.2021.1359.

Listrat, A. et al. (2015). How muscle structure and composition determine meat quality. *INRA Productions Animales, 28*(2), 125–136.

Maruska, K. P. (2001). Morphology of the mechanosensory lateral line system in elasmobranch fishes: Ecological and behavioral considerations. *Environmental Biology of Fishes, 60*(1), 47–75.

Maruska, K. P., & Tricas, T. C. (1998). Morphology of the mechanosensory lateral line system in the Atlantic stingray, Dasyatissabina: The mechanotactile hypothesis. *Journal of Morphology, 238*(1), 1–22. Retrieved from https://doi.org/10.1002/(SICI)1097-4687(199810)238:1.

Meredith, T. L., & Kajiura, S. M. (2010). Olfactory morphology and physiology of elasmobranchs. *Journal of Experimental Biology, 213*(20), 3449–3456.

Mills, M., Rasch, R., Siebeck, U. E., & Collin, S. P. (2011). Exogenous material in the inner ear of the adult Port Jackson shark, Heterodontus portusjacksoni (Elasmbranchii). *The Anatomical Record: Advances in Integrative Anatomy and Evolutionary Biology, 294*(3), 373–378.

Mogdans, J. (2019). Sensory ecology of the fish lateral-line system: Morphological and physiological adaptations for the perception of hydrodynamic stimuli. *Journal of Fish Biology, 95*(1 July), 53–72. https://doi.org/10.1111/jfb.13966.

Moss, S. A. (1972). The feeding mechanism of sharks of the family Carcharhinidae. *Journal of Zoology, 167*(4), 423–436.

Motta, P., Tricas, T., & Summers, R. (1997). Feeding mechanism and functional morphology of the jaws of the lemon shark Negaprion brevirostris (Chondrichthyes, Carcharhinidae). *Journal of Experimental Biology, 200*(21), 2765–2780.

Myrberg, A. A. (2001). The acoustical biology of elasmobranchs. In T. C. Tricas & S. H. Gruber (Eds.), *Behavior and sensory biology of elasmobranch fishes: An anthology in memory of Donald Richard Nelson* (pp. 31–46). Berlin: Springer.

Nelson, D. R., & Johnson, R. H. (1972). Acoustic attraction of pacific reef sharks: Effect of pulse intermittency and variability. *Comparative Biochemistry and Physiology Part A: Physiology, 42*(1), 85–95.

Nelson, D. R., & Johnson, R. H. (1976). *Some recent observations on acoustic attraction of Pacific reef sharks*. Long Beach, CA: California State University Long Beach Department of Biological Sciences.

Nelson, D. R., Johnson, R. H., & Waldrop, L. G. (1969). Responses in Bahamian sharks and groupers to low-frequency, pulsed sounds. *Bulletin of the Southern California Academy of Sciences, 68*(3), 131–137.

Newton, K. C., Gill, A. B., & Kajiura, S. M. (2019). Electroreception in marine fishes: Chondrichthyans. *Journal of Fish Biology, 95*(1), 135–154.

Oeffner, J., & Lauder, G. V. (2012). The hydrodynamic function of shark skin and two biomimetic applications. *Journal of Experimental Biology, 215*(5 March), 785–795. https://doi.org/10.1242/jeb.063040.

Patterson, J. C., Sepulveda, C. A., & Bernal, D. (2011). The vascular morphology and in vivo muscle temperatures of thresher sharks (Alopiidae). *Journal of Morphology, 272*(11), 1353–1364.

Peach, M. B., & Marshall, N. J. (2009). The comparative morphology of pit organs in elasmobranchs. *Journal of Morphology, 270*(6), 688–701. https://doi.org/10.1002/jmor.10715.

Pouca, C. V., & Brown, C. (2018). Food approach conditioning and discrimination learning using sound cues in benthic sharks. *Animal Cognition, 21*(4), 481–492.

Reif, W.-E. (1978). Protective and hydrodynamic function of the dermal skeleton of elasmobranchs. *Neues Jahrbuch für Geologie und Paläontologie, 157,* 133–141

Roa, J. N., & Tresguerres, M. (2017). Bicarbonate-stimulated soluble adenylyl cyclase is an acid-base sensor present in the cell cytoplasm and nucleus of elasmobranch fishes. *FASEB Journal, 31*(S1), 719.19–719.19. https://doi.org/10.1096/fasebj.31.1_supplement.719.19.

Schluessel, V., Bennett, M. B., Bleckmann, H., Blomberg, S., & Collin, S. P. (2008). Morphometric and ultrastructural comparison of the olfactory system in elasmobranchs: The significance of structure–function relationships based on phylogeny and ecology. *Journal of Morphology, 269*(11), 1365–1386. https://doi.org/10.1002/jmor.10661.

Shadwick, R. E., & Goldbogen, J. A. (2012). Muscle function and swimming in sharks. *Journal of Fish Biology, 80*(5), 1904–1939.

Silva-Garay, L., & Lowe, C. G. (2021). Effects of temperature and body-mass on the standard metabolic rates of the round stingray, Urobatis halleri (Cooper, 1863). *Journal of Experimental Marine Biology and Ecology, 540*, 151564.

Simonitis, L. E., & Marshall, C. D. (2022). Microstructure of the bonnethead Shark (Sphyrna tiburo) Olfactory Rosette. *Integrative Organismal Biology*, obac027. https://doi.org/10.1093/iob/obac027

Strasburg, D. W. (1963). The diet and dentition of Isistius Brasiliensis, with remarks on tooth replacement in other sharks. *Copeia, 1963*(1), 33–40. https://www.jstor.org/stable/1441272#metadata_info_tab_contents.

Sullivan, T., & Regan, F. (2011). The characterization, replication and testing of dermal denticles of Scyliorhinus canicula for physical mechanisms of biofouling prevention. *Institute of Physics, 4*(October). 046001. https://doi.org /10.1088/1748-3182/6/4/046001.

Syme, D. A., & Shadwick, R. E. (2011). Red muscle function in stiff-bodied swimmers: There and almost back again. *Philosophical Transactions of the Royal Society of London Series B, 366*(1570), 1507–1515.

Tapanila, L. et al. (2013). Jaws for a spiral-tooth whorl: CT images reveal novel adaptation and phylogeny in fossil Helicoprion. *Biology Letters, 9*(2 April), 20130057. https://doi.org/10.1098/rsbl.2013.0057.

Tester, A. L., & Kendall, J. I. (1969). Morphology of the lateralis canal system in the shark genus Carcharhinus. *Pacific Science, XXIII*, January 196, 1–16.

Thewissen, J. G. M., & Nummela, S. (2008). *Sensory evolution on the threshold: Adaptations in secondarily aquatic vertebrates.* Berkeley, CA: University of California Press.

Tomita, T., Murakumo, K., Komoto, S., Dove, A., Kino, M., Miyamoto, K., & Toda, M. (2020). Armored eyes of the whale shark. *PLOS ONE, 15*(6), e0235342.

Van den Berg, A. V., & Schuijf, A. (1983). Discrimination of sounds based on the phase difference between particle motion and acoustic pressure in the shark Chiloscyllium Griseum. *Proceedings of the Royal Society of London. Series B, Biological Sciences, 218*(1210), 127–134.

Watanabe, Y. Y., Lydersen, C., Fisk, A. T., & Kovacs, K. M. (2012). The slowest fish: Swim speed and tail-beat frequency of Greenland sharks. *Journal of Experimental Marine Biology and Ecology, 426*, 5–11.

Wilga, C. D. (2005). Morphology and evolution of the jaw suspension in lamniform sharks. *Journal of Morphology, 265*(1), 102–119.

Wilga, C., & Motta, P. (1998). Conservation and variation in the feeding mechanism of the spiny dogfish Squalus acanthias. *Journal of Experimental Biology, 201*(9), 1345–1358.

Wilga, C. D., & Motta, P. J. (1998). Feeding mechanism of the Atlantic guitarfish Rhinobatos lentiginosus: Modulation of kinematic and motor activity. *Journal of Experimental Biology, 201*(23), 3167–3183.

Yopak, K. E., Lisney, T. J., & Collin, S. P. (2015). Not all sharks are 'swimming noses': Variation in olfactory bulb size in cartilaginous fishes. *Brain Structure and Function, 220*(2), 1127–1143.

Zhai, Y., Zheng, X., & Xie, G. (2021). Fish lateral line inspired flow sensors and flow-aided control: A review. *Journal of Bionic Engineering, 18*(2), 264–291.

4

Shark conservation and fisheries management across the globe

Written by Triana Arguedas Alvarez and
Camila Cáceres with contributions from Lauren
Ali, Sara Asadi Gharbaghi, Lara Fola-Matthews,
Ingrid Hyrycena dos Santos, Devanshi Kasana,
Buddhi Maheshika Pathirana, Ilse Martinez Candelas,
Ana P.B. Martins, and Angelina Peña-Puch

DOI: 10.1201/9781003260370-4

Triana Arguedas Alvarez

 I was born in Ohio to two Mexican parents that recently immigrated to the United States. My family grew, and we moved to South Florida, to Boca Raton which I consider my hometown. I have always loved the outdoors and became interested in answering questions about how the world works. My first encounter with this newfound curiosity was learning about photosynthesis. I continued my pursuit of learning about the natural world at the University of North Florida, where I earned my degree in Interdisciplinary Studies titled, "Marine Communities and Conservation" with a minor in Environmental Science. My capstone project focused on shark nursery grounds and possible trends of abundance in certain shark populations in Northeast Florida and Southern Georgia. During my time at UNF, I was involved in several sampling projects ranging from eel spawning in the GTM-NERR, oyster reef health, dolphin relationship sightings, and porcelain crab competition. Even with this exposure to the different ongoing projects at the university, sharks and rays became my passion. After taking a "Coastal Fisheries Management" course at UNF, the discipline of fisheries science interested me, which led me to explore the intersection of sharks and fisheries. Much to my surprise, I soon became engulfed in the world of global fisheries, specifically data-deficient fisheries. In the search for the next step in my academic career, I was able to connect with several shark scientists across the globe and eventually began research in the Yucatán Peninsula. Although I grew up spending time in Mexico City (where my family is from), moving to the Peninsula opened my eyes to the diversity and the rich culture of fishing communities.

My project focused on historic reconstructions and learning more about data-deficient fishery dynamics in the Mexican Atlantic. I am very involved with several organizations that offer mentorship programs for undergraduate minority students and first-generation students. I am passionate about marine ecology, sharks, and creating space for other minority students.

I currently work as a Biological Scientist in the Fish and Wildlife Research Institute, under the Fish and Wildlife Commission, in the Fishery Independent Monitoring Program.

Camila Cáceres

 I was born in Bogotá, Colombia, to a Colombian father and an Argentinian mother. By the age of five I decided I wanted to become a marine biologist after seeing the ocean for the first time in San Andres, Colombia. I immigrated to the United States when I was ten years old and soon after, in front of a shark tank at an aquarium in Charlotte, North Carolina, I realized that I wanted to dedicate my life to shark research. I went on to pursue my passion, and in 2012 received a degree in Biology (BSc) from Duke University. Upon completing my studies, I was hired as a research assistant at Stanford University's Hopkins Marine Station, where I was first introduced to fisheries research, a topic that would change my path forever. Armed with a love for sharks and an urge to learn about small-scale and sustainable fisheries, I completed my MSc and PhD in the Heithaus Lab at Florida International University.

My research focuses on gathering baseline data on coastal small-scale fishing communities and coral reef sharks and rays, in the Caribbean Sea. I was able to sample and collaborate with scientists in Colombia, Guadeloupe, Martinique, Tobago, and Florida Keys. I have met with Colombia's Vice-President and various diplomats to discuss marine research, have been featured in Discovery, National Geographic, and Telemundo media among others, and was given the Professional Award at the academic conference Sharks International for my research in the Caribbean. As an immigrant, Latina, and LGTBQ, I actively promote diversity and inclusion in STEM and the outdoors.

Lauren Ali

I am incredibly fortunate to come from Trinidad and Tobago in the Caribbean, a place whose phenomenal biological and cultural richness has shaped my perspectives both as a scientist and as an inhabitant of the Earth. In this diverse culture which puts a high value on education there was no shortage of female role models of all races and ethnicities, but like those who came before me, I had to grapple with the national and regional colonial legacies that have negatively impacted investment in local capacity for biological science. Today, I am proud to be a member of the country's small but strong and highly active scientific community. Although it's not the easiest, I appreciate my research environment because I believe it's a situation where driven individuals can accomplish a lot for underserved populations, both human and non-human.

In Trinidad and Tobago shark meat is a cultural food, but their shark fishery is one of many around the world that is data-deficient and where data collection agencies are under-resourced. The artisanal fishery lands over 30 shark species, but sadly, elasmobranchs have almost no legal protections or fishing regulations (even the endangered ones like the scalloped hammerhead). In the quest to balance shark needs and human needs, I studied Trinidad and Tobago's artisanal shark fishery and the ways that people perceive sharks.

My MPhil research at the University of the West Indies involved both biological and social science methods. I studied the biology of two species of artisanally fished Rhizoprionodon sharks as influenced by gear selection for size and maturity, as well as interviewing fishers and restaurants about their knowledge and practices regarding sharks, and interviewing members of the public to understand shark consumption and how people perceive sharks and shark conservation in a culture where they are mainly considered food.

My work provided valuable insights into the demand for shark which drives unsustainable exploitation through a supposed bycatch fishery, and highlighted inconsistencies between public attitudes and practices which can hopefully be the subject of further research aimed at promoting behavior change.

In my non-shark time, I also work studying the illegal wildlife trade in terrestrial animals like capuchins, finches, and parrots with the aim to improve animal welfare.

Sara Asadi Gharbaghi

 My name is Sara Asadi Gharbaghi. I studied marine biology and received a BSc then a MSc in marine ecology. Since the second year of my BSc, I have been interested in studying the behavior of marine animals. For my MSc I studied the behavior of cuttlefish in the Persian Gulf to understand which camouflage tactics they choose in facing different stress factors and predators in different areas. After days of scuba diving and field study my soul was touched by the behavior of their predators as well and I became interested in studying why and where I could find different types of sharks and rays. I was aware of how much harder it is to see them in the wild, however. I could find them in street food menus around the beach much more easily than I could underwater in my area in the Persian Gulf.

Day by day I became more interested in how I could help build a more sustainable shark fishery and it became my main goal in life to find ways to protect sharks by learning more about them and discovering their mysterious world. Unfortunately, in Iran, there was no specialized program to study or learn about sharks. So, I had to move abroad where I could discover their world better. I chose the Philippines and participated in an internship program (LAMAVE) as a research assistant where I had the chance to learn about sharks and use BRUV systems to identify them and tag them to study their behavior and migration patterns. I was also able to take samples of different species, including the deep-sea sharks that had always fascinated me. After spending several months in the Philippines, I headed to Indonesia to complete a volunteer job identifying reef sharks on GILI island.

After returning to Iran, I founded my NGO under the name of "Raysha Persia" in 2017 which started to educate local people in southern Iran which borders the Persian Gulf to stop fin fishing. Over the past four years, I focused my research on shark ecology. I spend my days in the field sampling, scuba diving, tagging, using BRUVs, and analyzing data. I try hard to improve my abilities and knowledge. This year, with the help of MISS, I have begun doing new research on whale sharks in the Persian Gulf.

Lara Fola-Matthews

I am a Nigerian scientist currently living in Nigeria. Growing up, I had the privilege of enjoying both academic and social activities as there are lots of beautiful beaches in Lagos. My parents are beach lovers as well, so most of my holidays were spent with visits to the beach. My siblings and I had a pet fish and would pick seashells, cowries, and other washed-up ornaments to decorate its bowl. My father worked for a fishing company at some time in his life and he had a sailor friend who gave me the chance to watch the fishers offload their landings from the fishing vessels whenever I visited my dad. I was always fascinated by the megafauna landed, sharks and rays included. However, at that time I never knew there were different species of sharks because they all looked the same to me. I also never would have guessed I would be a shark scientist and conservationist later on in my life. I graduated from the Department of Marine Sciences, the University of Lagos for my BSc. Years later, I gained admission to the same university for an MSc in Fisheries Biology and Management. I am currently a PhD student at the same University with a research focus on shark biology, ecology, and conservation. I am presently studying the Bio-Ecology, species composition, and species identification of sharks off the Nigerian Coast using DNA Barcoding methods. I intend to collect data to understand the diet, growth pattern, and reproductive biology of these sharks and understand the chemistry of their environment. This will represent a baseline study in Nigeria as there have not been many scientific studies of sharks from the Nigerian waters published online. Another interesting part of my research is the use of *local ecological knowledge (LEK)* to understand the trend in shark fisheries in past years in comparison to the results that would be obtained from my study. I record data of landings from both the artisanal fishermen and the industrial fishing vessels to investigate the composition of the catch and its abundance. I identify the species of sharks with the aid of identification guides. I also take fin clips during my trips for the purpose of DNA barcoding which will serve as a more accurate means of identification. The molecular aspect of my study is funded by the MISS Rising Tide Mentorship Program. I'm also a steward for the blue economy. I spend my free time educating young learners and coastal dwellers on the importance of sharks in the ecosystem in a bid to mitigate the negative perceptions they might have of sharks. I have high hopes that the findings from my research and advocacy will assist ongoing efforts to develop a plan of action for shark fisheries management in Nigeria.

Ingrid Hyrycena dos Santos

I was born in a city named Goiânia, in the Midwest of Brazil but my family moved to Balneário Camboriú, where I grew up. Both of my parents worked at an environmental organization of the government and because of that I grew up surrounded by scientists, with one particular oceanographer being special to me. He is a well-known elasmobranch specialist and had the patience to teach a five-year-old kid everything he could about sharks and rays. The curiosity for these misunderstood ocean creatures soon turned into a deeper passion. Since then, I have dedicated my entire life to becoming a scientist and, in 2015, I started my undergraduate course in Biological Sciences at Universidade do Vale do Itajaí. In July 2021, after six years of university, I became officially a biologist.

For four and a half years, I worked as a volunteer intern at the Aquatic Ecosystems and Fisheries Laboratory at UNIVALI, developing research focused on the biological aspects (reproduction and feeding habits) of deep-sea and coastal sharks and rays incidentally caught by both industrial and artisanal fisheries. In March 2019, I was an intern at Oceans Research in Mossel Bay (South Africa) and had the chance to participate in different projects involving marine life, especially about great white sharks. I founded the "Shark Project," which seeks to spread scientific information, for free, using more accessible language on social media. In 2020, I started a project with artisanal fishers aiming to characterize the beach seine and study the ichthyofauna of Balneário Camboriú. At the same time, I consolidated a compensatory release and monitoring program for cownose rays and other threatened species that were incidentally caught.

As a result of this partnership with fishers, we released more than 750 stingrays in less than two years. In the beginning of 2022, I was approved in the first place of the master's degree selective process and started studying at the Federal University of São Paulo (UNIFESP). Now I am part of the Fisheries Science Laboratory (LabPesca/UNIFESP), where my research is focused on the age and growth of two species of rays that occur in Brazil.

Devanshi Kasana

I grew up in a landlocked desert state in India and ironically developed an inclination to explore far-removed landscapes that I felt an interesting sense of kinship with. My curious love for all things marine nurtured by my parents, eventually transformed into a career. What drew me to conservation was the complexity of challenges encountered when managing people and wildlife simultaneously. Choosing an unconventional career choice came with its own set of challenges—navigating academia, finding relevant work opportunities, parental apprehensions, sociocultural norms; yet the sense of purpose and fulfillment derived from pursuing this has been unparalleled. I started out with a BSc in Life Sciences at Miranda House, University of Delhi and went on to complete a MSc in Conservation at University College London. Shortly after, I took an opportunity to work with the World Wide Fund for Nature India which enabled me to get hands-on experience through various marine and coastal projects in India. A brief stint as a research assistant for a global shark survey drew me back to academia and prompted a move to Miami to pursue a PhD in Biological Sciences at Florida International University.

I am especially interested in using an interdisciplinary approach to understanding shark fisheries, consumption, and conservation. I use tools such as genetics, ecotoxicology, and sociology to guide consumer awareness, education, and behavior change. This is increasingly relevant in developing nations where the discourse around sustainability must balance livelihood demands and environmental protection. My current work focuses on investigating the fishery that supplies shark meat and its trade dynamics in Belize, Central America. My research looks at how shark meat moves between the fishery and markets, and is designed to identify factors that influence the demand and supply of this commodity. The aim is to gather data that informs community-engaged conservation models. I am also exploring the presence of heavy metals in shark meat from a food safety context. I envision being able to use my current research to inform policy-level decision-making processes around shark fisheries management in Belize. I enjoy being able to look at conservation issues from a comprehensive lens that views humans and wildlife as aspects of nature that must coexist in harmony; and believe that successful, impactful conservation interventions are guided by science but enacted by people, and therefore must be inclusive and equitable.

Buddhi Maheshika Pathirana

I was born in a rural village in Sri Lanka known as Erathna which is situated at the bottom of Adam's peak and, about 150 km away from the country's capital, Colombo. I was lucky to become the first student from our village to graduate from the Ocean University of Sri Lanka in the Fisheries and Marine Sciences program. After graduation, it was quite a challenging beginning for me, especially to find a suitable research organization in Sri Lanka. Only a few organizations are there and only one or two are willing to support female undergraduate students. I was finally fortunate to join a marine research and consultant organization here in Sri Lanka, known as Blue Resources Trust (BRT), which enabled me to conduct my final year thesis project as a part of their Sri Lanka Elasmobranch Project. There I focused on assessing the ray landings in Sri Lankan harbors, markets, and landing sites. After finishing my degree, BRT identified my potential as a young female enthusiastic researcher and offered me a job to work as a permanent researcher on the project. And now, to broaden my opportunities, I have taken up the position of Administrative Officer with Blue Resources Trust.

Since 2017, I have been conducting fisheries surveys on the west, northwest, south, and east coast of Sri Lanka to determine shark and ray biodiversity. I have spent more than 1,000 days at harbors, markets, and landing sites to fill in information gaps present in the elasmobranch captures from artisanal and small-scale fisheries in the region. The baseline data collected by me includes identification photos, measurements, sex, maturity, fishing gear used, and catch locations, which are all vital parameters when it comes to the conservation and management of these species. Tissue samples that I collected will also be used for genetic population assessments and stable isotope analysis. After each survey day, the main goal is to identify each specimen up to species level and enter all collected data into our national database within 24 hours. Apart from this general fishery survey, I have conducted 100 fisheries dependent questionnaires on sawfish, guitarfish, and wedgefish as part of the larger Indian Ocean study, which was a joint collaboration between Blue Resources Trust and the Elasmo Project.

Ilse Martinez Candelas

 I spent the first part of my childhood in Guerrero, one of the poorest states in Mexico, where I witnessed the inequality of rural Mexico. My family later moved to Mexico City but seeing the poverty of rural communities during my childhood defined my perspective of the world and my wish to make a change. I always loved the ocean, so I studied Hydrobiology. Despite having a great time going through college and learning about marine life, I felt that something was missing. Trying to find my academic passions, I did an internship to study sharks, and later on, volunteered for a coral reef monitoring project in the Mexican Caribbean. This last project allowed me to get to know different fishers, where I discovered that I loved listening and learning from local people. Afterward, I did my master's at El Colegio de la Frontera Sur, where I combined my interest in sharks and my wish to learn from fishers. Thanks to my amazing supervisor Iván Méndez, I learned about the existence of historical ecology and realized that it combined everything I loved. My fieldwork has given me the opportunity of getting to know a lot of fishers in the Yucatán Peninsula, especially in Campeche. So now you may find me in the coastal communities of southern Mexico interviewing fishers while we share a good laugh and maybe a cold soda if it is hot.

My research focuses on the use of local ecological knowledge to understand the history and development of artisanal elasmobranch fisheries in Campeche and the Yucatán Peninsula. I tend to use different information sources to reconstruct the fishing history of the different coastal communities. The ultimate goal of my research is to generate social and biological information that may help to improve fisheries management. The long history of exploitation in the southern Gulf of Mexico has led the youngest generation of fishers to ignore the past diversity and abundance of elasmobranchs in the area. My latest project has two objectives; the first one is to establish baseline information about the ecology and distribution of the elasmobranch species in Terminos Lagoon, Campeche, through literature review and interviews. The second one is to transfer this knowledge to the youngest generation of fishers through a short documentary starring local elder and middle-aged elasmobranch fishers, and through the use of participatory mapping exercises. The aim is to generate intergenerational dialogues that bring the communities together and inform them about the available management tools. I believe it is important to empower and include local people and their local history during the development of management or conservation strategies for these to be successful and ethical. Conservation needs local leaders; we are just giving them the tools to take ownership of their resources. You can check the details of my project "Mexico's forgotten sharks: rediscovering a natural legacy" on the Save Our Seas Foundation webpage.

Ana P.B. Martins

I was born and raised in São Luís, an island in northeastern Brazil. My fascination for the underwater world began at an early age while playing in Amazon rivers and on beaches of the Brazilian shores. My interest in marine science grew with time and, since elementary school, I started to plan on how to pursue a career working on my favorite animal, sharks. I completed a BSc in Biological Sciences and an MSc in Biodiversity and Conservation at the Universidade Federal do Maranhão, studying the supply chain and conservation status of sharks based on traditional fisher knowledge. I then decided to take a much bigger step and moved to Australia to pursue a PhD at James Cook University. There, I developed a project that integrated biotelemetry approaches (using small tracking devices to learn about shark movements) and stable isotope analysis (collecting small tissue and blood samples from sharks to find out about their food webs). My goal with this project was to refine the ecological roles of juvenile stingrays in coral reef ecosystems––a topic that at the time had received very little attention. Currently, I am a postdoctoral fellow at Dalhousie University in Canada working on a multiyear project that aims to unravel the global shark meat trade.

I am mostly driven to work in areas of spatial and trophic ecology, human dimensions of fisheries resources, and sustainability and conservation, especially in developing nations with multifaceted socioeconomic and cultural structures. I believe that by doing this work, I will be able to help improve our knowledge of shark ecology and fisheries, and hopefully provide useful information for the development of effective management and conservation strategies.

Angelina Peña-Puch

 I was born and raised in the city of San Francisco de Campeche, Campeche, Mexico. Since I was a child I did several tours along the coast of the Yucatán Peninsula. In the company of my parents I enjoyed the coastal-marine landscapes, the beaches, small-scale boat rides, and the gastronomy made with local fish and seafood. My hometown stands out as a Cultural Heritage of Humanity for its buildings and traditions.

While studying for a degree in Economics at the Autonomous University of Campeche (UAC) I had the opportunity to do my professional internship at the Institute of Ecology, Fisheries and Oceanography of the Gulf of Mexico (EPOMEX). There I learned about the importance of marine resources, the coastal zone, and the mangrove habitats of the Gulf of Mexico. My interest in the protected coastal areas of Campeche grew. These areas include the Celestún Biosphere Reserve, the Petenes Biosphere Reserve, and the Flora and Fauna Protection Area of the Laguna de Términos. As a result of this stay, a scientific communication article was published (Espinoza-Tenorio, et al., 2012).

Between 2016 and 2020 I participated in several professional development workshops including "Fisheries Management with an ecosystem approach," "Social Management for the Protection of the Environment and Territory" of FASOL-ECOSUR, "Introduction to Rights-Based Fisheries Management" by Environmental Defense Fund of Mexico (EDF of Mexico), and a student Workshop on Management of International Coastal and Marine Environments in the Gulf of Mexico by Harte Research Institute for Gulf of Mexico Studies at Texas A&M University Corpus Christi.

In 2019 I taught a course on Sustainable Management of Aquatic Resources for the Master of Science in Natural Resources and Rural Development at El Colegio de la Frontera Sur (ECOSUR). Additionally, I led a course titled, "Use of government information in Mexico for fisheries databases" at ECOSUR. I received my PhD in Ecology and Sustainable Development, from ECOSUR with the thesis "Evaluation of the sustainability of the socio-ecological fishing systems of Campeche." Between 2020 and 2021 I published three scientific articles: "The sustainability of *small-scale fisheries* in oil-producing sections of the Gulf of Mexico"; "*Socio-ecological systems* as a management unit: the case of the fisheries of Campeche, Mexico"; and "Advances in the study of Mexican fisheries with the social-ecological system (SES) perspective and its inclusion in fishery management policy." My research interests include sustainability sciences, natural resource management, socio-ecological fisheries systems, fisheries policy instruments, and fisheries sustainability standards. I am now headed into my postdoc at the Universidad Autónoma de Campeche (UACam).

KEY TERMS

Exclusive Economic Zone (EEZ): the area of the sea in which a sovereign state has special rights regarding the exploration and use of marine resources, including energy production from water and wind

fecundity: the ability to produce an abundance of offspring or new growth; fertility

FAO: the United Nations Food and Agriculture Organization, which plays a leading role in international fisheries management and policy

Knowledge Attitudes and Practices (KAP): a framework used to conduct representative studies on specific populations which focuses on investigating what humans know and feel about a topic as well as their associated actions

Socio-Ecological Systems (SES): consideration of humans in different ecosystems and the extraction of services and products from the environment

subsistence fishing: fishing activity specifically for the direct, personal consumption of marine products without larger commercialization

small-scale fisheries: smaller size and capacity vessels that focus on shorter trips, less human power to run trips, smaller management teams, and is usually associated with local consumption of the product

traceability: the ability to access information about a product and its history through the whole, or part, of a production chain. It answers the question of "when and where the product was produced by whom"

trophic levels: hierarchical groups which organisms belong to, determined by their feeding habits and preferences

INTRODUCTION

Overfishing has been a local and regional problem for hundreds of years, and recently has become a global challenge. In addition to collapses of traditionally targeted species (Bundy, 2005; Fromentin et al., 2014; Swain & Benoit, 2015), populations of large-bodied marine predators, including marine mammals and sharks, have been quickly declining on a global scale (Heithaus et al., 2008). Although humans have historically preferred fish species near the top of food webs (Pauly et al., 2005; Sethi et al., 2010), fisheries now target species at many different *trophic levels* (Essington et al., 2006). Elasmobranchs are mid and upper trophic level predators in oceans worldwide (Cortes, 1999). They have been harvested around the world by industrial, artisanal, and recreational fisheries, and they are landed with a variety of fishing gears and vessel types (Prince, 2002; Musick & Bonfil, 2005). Assessing elasmobranch fisheries has proven to be difficult in many situations due to lack of species-specific data, a lack of data on population structure, and the highly migratory nature of many species (Calich et al., 2018).

In addition, many sharks are taken illegally and/or catches are not reported (Worm et al., 2013). Sharks are more susceptible to the effects of fishing compared to bony fishes because of their low *fecundity*, slow growth and late maturity (Firsk et al., 2001; Mollet & Cailliet, 2002; Cailliet, 2015). These life history characteristics combined with heavy fisheries pressure have led to significant declines in elasmobranch populations in coastal, reef-associated and pelagic ecosystems (Baum et al., 2003; Dulvy et al., 2008: Ferreti et al., 2010). Currently there are over 260 (about 25%) elasmobranch species listed as Vulnerable, Endangered, or Critically Endangered on the IUCN Red List (Dulvy et al., 2014; IUCN, 2021).

This chapter will review various current elasmobranch fisheries topics including *small-scale fisheries*, methods to study shark fisheries, shark meat, and fin trade, and the *socio-ecological* importance of shark catches.

THE PAST

SHARK MEAT: FROM THE BOAT TO OUR PLATES

Shark meat is an important source of both food and income in most coastal areas of the world (Baker-Médard & Faber, 2020). There is evidence of shark meat consumption as early as the Chalcolithic and Bronze Age (Mojetta et al., 2018), with more consistent reports starting in the fourth century (Nuñez, 2007). In the early 20th century, shark meat was popular with communities in coastal areas where meat could be consumed fresh. Conversely, it was considered unfavorable as food in many nations due to the lack of proper refrigeration or ice and the improper handling of sharks which resulted in a strong odor and taste. However, this changed with the introduction of refrigeration, which facilitated better handling of shark meat (Vannuccini, 1999), as well as with the start of large-scale commercial exploitation of sharks after the First World War (1914–1918) (Rose, 1996). Nowadays, shark meat markets continue to expand as a result of regulations encouraging the full utilization of sharks (e.g. anti-finning laws), increased demand for protein sources, and the decline of commercially valuable fish species (Dent & Clarke, 2015). Consequently, shark meat––previously considered unpalatable––is now being traded and consumed globally. In fact, in the past few decades, the total value of shark meat reached USD 2.6 billion, surpassing the trade in fins by both volume and value (Niedermüller et al., 2021). As demand for shark meat continues to rise, it becomes crucial to further understand the trade and fisheries supplying this market at global and regional scales (Figure 4.1).

A growing market creates trade networks that enable products to move "boat to plate," i.e., from a fishery to the consumer. This is done through a supply chain which includes all the different stages of creating and selling a product to the consumer. In fisheries these supply chains consist of: fishing, landing/processing, sales, transport, and consumption (United Nations, 2019). Transparency and *traceability* throughout the supply chain help ensure that trade is legal, protected species remain off

Figure 4.1 Sphyrna tiburo *(bonnethead shark) landings from a coastal longline fishery in Belize.*

the market, and consumers are able to make informed consumer choices. Global shark meat markets are diverse and geographically dispersed, with supply chains that include several domestic and international actors interacting with each other at various steps of the process (Dent & Clarke, 2015). For example, Brazil and Spain are the top global importers of shark meat by volume. Brazil meets its shark meat demand by supplementing domestic shark catch with imported volumes. On the other hand, Spain supplies its domestic demand through imports and domestic production while also being the world's largest exporter of shark meat. In such a case, there is no one dominant centre of activity that can be studied, rather a comprehensive understanding of global players is required for a true representation of the trade (Dent & Clarke, 2015). This includes not only major importers and exporters, but also trade facilitators playing a pivotal role in connecting different trade partners in the network (Niedermüller et al., 2021) (Figure 4.2).

Shark meat markets are particularly difficult to trace due to the lack of accurate catch and trade data, as well as the common practices of meat mislabelling (Hobbs et al., 2019). Fisheries data often covers only a proportion of the sharks caught and the traded by-products such as meat and fins. This is because catch statistics are rarely distinguished between species; and trade records do not allow accurate identification

Figure 4.2 *Multispecies catch of deepwater sharks in Belize.*

of species or the tracking of catch volumes or economic values over time. Mislabeling further complicates these challenges and it can occur both unintentionally due to misidentification of species, or intentionally to substitute depleted species, increase profit margins (Marchetti et al., 2020), and supply the market demand for these products. Mislabeling ultimately results in uninformed decision-making by consumers, potentially exposing them to higher levels of heavy metals such as mercury found in fish, also making it a food safety issue (Bornatowski et al., 2015). Therefore, governments and the seafood industry must encourage accurate labeling practices and supply chain monitoring as an ethical responsibility to the public (Almerón-Souza et al., 2018). In the meantime, DNA analysis techniques can be used as a tool for species identification to detect and monitor mislabeling in seafood products.

Researchers have only begun to decipher the shark meat trade dynamics by identifying important trade routes, key import, export and re-export nations, and trade flows by volume and value (Dent & Clarke, 2015; Braccini et al., 2020). Additionally, the lack of information on drivers and patterns of shark meat consumption around the globe has arisen as a key knowledge gap. Although recent studies shed light on the need to understand domestic shark meat consumption and trade (López de la Lama et al., 2018; Karnad et al., 2020; Ali et al., 2020; Seidu et al., 2022), more studies encompassing different cultures and geographic locations are required for a comprehensive understanding of the global patterns of shark meat consumption.

These new techniques are important because sharks are crucially important for maintaining ecosystem functioning in marine food webs and their loss can have implications for the environment and the economic activities dependent on their continued existence. Additionally, shark meat plays an important role in food and nutritional security; the fishery and trade generate jobs and income for the various stakeholders.

A good way for people to be informed consumers is to be curious about what they put on their plates. Encouraging transparency in markets can begin with holding restaurants, fishmongers, and retailers accountable for having a sustainable sourcing policy. This means asking where and how a fish was caught and what species it is. Local sustainable seafood guides provide information to help identify whether products from these species are sustainable. Sharks and rays are often sold under generic names that disguise several species, which means that a clear answer might not always be available. In situations where shark products are not being used to meet protein demands or when product traceability cannot be confirmed, the safest solution is to avoid consumption.

USING HISTORIC DATA TO GIVE INSIGHTS INTO FUTURE FISHERIES

Humans are deeply connected to the ocean; some of the first civilizations (Aztecs and Olmecs) established their most important cities close to the sea. Specifically, sharks have been an important food source for centuries in many different cultures and have an important economic and cultural importance. In Mexico, for example, shark and ray remains were found in different Mayan pyramids in what is now the Yucatán Peninsula. During the Aztec empire, Veracruz, a coastal state, had important port cities that frequently transported shark meat to the capital city of Tenochtitlan, now Mexico City.

This importance did not disappear over time; coastal communities continued to exploit different shark populations. The only difference is the active effort with which this dramatic exploitation occurs (Bonfil, 1997; Pérez-Jiménez and Méndez-Loeza, 2015). There is a big difference between landing sharks with a canoe (Figure 4.3) versus modern boats (Figure 4.4) (Martinez-Candelas et al., 2020). At the start of formal commercialization, even at the local level, boats and fishing equipment were constructed out of natural materials. These vessels could only make small trips since they couldn't support large quantities of landings. Once fishing technology began to advance, these staples of coastal fishing communities were soon replaced by higher capacity motors, on board ice chests, and GPS systems. Larger vessels made from sturdy fiberglass provided fishers with the opportunity to go on extended trips and process larger landings without the risk of damaging their smaller vessel (Martinez-Candelas et al., 2020).

The increase in vessels over time generated some of the highest historic landings in Mexican history (32, 576 tons in 1995) (Castillo-Géniz et al., 1998), moving Mexico within the top ten shark and ray producers in the

Figure 4.3 *Typical fishing canoe with sails attached, generally equipped with fishing gear and only one or two men.*

Figure 4.4 *Popular fishing vessels now powered with an onboard motor, equipped with GPS, interchangeable gear types, and larger capacity for more fishers.*

world (Castillo-Géniz et al., 1998). The cultural importance and access to marine resources are seen today, with some of the largest fish markets in the country distributing Mexican-caught shark products across the nation and overseas.

Quantifying shark fisheries can be difficult, but with the use of historic records and other research methods, historic and current landings can be compared along with changes in the fishing sector as the economy changes. Combining different information sources such as surveying different generations of fishers, reviewing old reports, checking old books that describe fisheries in the areas, gathering early research papers published for different study areas, and some people have gone as far as analyzing old cooking recipes to learn which species were common in the past.

An example of how history gets used in fisheries studies is seen in a sector of fisheries science called "data-deficient" fisheries. This term is used to describe fisheries that do not use species specific information to manage their stocks. Generally, artisanal and small-scale fisheries fall into this category since data collection comes in the form of permit holder logbooks, which are susceptible to underreporting. Keeping this in mind, there are ways to supplement the uncertainty associated with underreporting which usually come from personal accounts such as interviews and comparisons of data gathered through literature. Using vulnerability studies, which incorporate the social aspect of fishers and the life histories of different species of sharks, and historic reconstructions, patterns may emerge from history to tell the story of these stocks across time.

It is necessary to analyze the different fisheries from a historical perspective. With information available, it's easy to think that all shark fisheries are at the peak of their exploitation, which may not be the case. In the southern Gulf of Mexico, the largest landings occurred during the 1990s and early 2000s, and shark fisheries rarely catch large sharks anymore as it has become unprofitable to spend time and resources looking for these rare sharks (Pérez-Jiménez & Méndez-Loeza, 2015).

Why is the past important? It can help scientists understand the cultural context in which the different shark fisheries have developed, providing important information to develop successful management strategies. The increased involvement of fishers and permit holders has also become a key factor in the development of effective fishery planning, since it considers the livelihood of coastal communities while also trying to assure future generations can participate in fishing activities. New and innovative ways of studying data-deficient fisheries have already provided strategies for fishery managers to begin shifting efforts in different ways to create sustainable management.

THE IMPORTANCE OF SOCIO-ECOLOGICAL SYSTEMS IN ELASMOBRANCH FISHERIES FOR THE STATES OF THE YUCATÁN PENINSULA

Declines in fishery productivity is a problem that has severely impacted poorer, more underdeveloped coastal regions. As a consequence of the human–environment relationship and the use of blanket fisheries management measures, which have resulted in unsustainable fisheries, the need to improve fisheries management is based on the combined analysis of social and ecological systems and the integration of local knowledge. Fisheries are now recognized as *socio-ecological systems (SES)*, where fishers are acknowledged as part of the marine environment. However, the most widely used management unit in fisheries is limited to the target species and their associated species, without integrating fishers in formal decision-making.

In Mexico, most of the fishing effort is made up of small-scale vessels measuring less than 10.5 m in length, with or without an outboard engine, with limited autonomy for trips, and space for an ice storage system (DOF, 2007). This fishing activity is focused on several bony fish species, where their market value influences the equipment used and length in trips. Fishing cooperatives, or fisher unions, in these communities have permits for a variety of bony fish species, usually referred to as a "multispecific" fishery. Since fishing resources vary by abundance, availability, seasonal closures, and price changes, these considerations have to be taken into account when establishing regulations (Fernández et al., 2011) (Figure 4.5).

In Campeche, small-scale shark fisheries have been recognized as part of multispecific bony fish fisheries (e.g. Snapper) and operate

Figure 4.5 *Current small-scale and artisanal boats with an onboard motor, ice chest, and various gear types used along the peninsula.*

based on species availability and seasonal closures (Pérez-Jiménez et al., 2016). Although there are two size categories (large and small) in Mexican fisheries, they are treated as a single group and do not have species-specific management plans that consider different biological characteristics that make certain species either more or less vulnerable to fishing pressures. Much of the large and small sharks landed are consumed locally, since it is a source of food in the Yucatán Peninsula. Once the landings presented the opportunity to consume a cheap source of protein, cultural dishes made from shark and ray precuts became staples of the region.

In the northern area of Campeche (Lerma, San Francisco de Campeche, and Isla Arena), the small shark fishery endures high fishing pressures. Among the most popular species include the tutzún (*Rhizoprionodon terraenovae*; sharpnose shark) and pech (*Sphyrna tiburo*; bonnethead shark), usually caught with gillnets. This region has large mangrove areas as part of the Celestun Biospheric Reserve and the Petenes Biospheric Reserve. In comparison, in the central and southern coast of Campeche, fishing for small sharks is notably less common compared to the north. Some stingray species of commercial importance are the balá (*Hypanus americanus*; southern stingray), pinta (*Aetobatus narinari*; spotted eagle ray), and chucha (*Rhinoptera brasiliensis*; cownose ray). These species are generally caught with a combination of gillnets and longlines. The change in gear creates a problem in management since it is not possible to count how many fishers use specific types of gear. This is an example of integrating fisher knowledge and creating relationships between fishery managers, permit holders, and lawmakers.

Yucatán has the unique ecosystem feature of freshwater springs off the coast which fosters the small shark fishery, specifically for *R. terraenovae*, caught most commonly by gillnet. In the central area of the coast the fishing intensity is moderate, but also incorporates the use of longlines. On the eastern coast in the villages of San Felipe and Rio Lagartos there is low shark fishing pressure since these villages mostly contribute to the lucrative regional lobster fishery (Pérez-Jiménez et al., 2016).

As the coast wraps around into the Caribbean Sea, the state of Quintana Roo has a history of shark fishing activity. While the state was mostly unknown to people outside of Mexico, once cities like Cancún and Playa del Carmen developed, the state became a tourist hub. Now, once frequently caught shark species such as the toro (*Carcharhinus leucas*; bull shark) became a tourist attraction, where diving tours allow visitors to swim with these animals along with other species such as the *tiburón ballena* (*Rhinchodon typus*; whale shark). Some species are protected by law, but others are not and are vulnerable to overexploitation and conflicts of interest. For example, the city of Isla Mujeres was a *C. leucas* landing capital of the peninsula although conflicts between managers in shark ecotourism and fishers have not been able to establish reasonable parameters for either party to flourish. The state of Quintana Roo no longer identifies as a shark fishing state, but rather a shark tourism state which has led to decreases in shark fishing efforts (Rubio-Cisneros et al., 2019; Blanco-Parra et al., 2016). Some species of shark that are still landed in the state include the tigre (*Galeocerdo cuvier*; tiger shark), gata (*Ginglymostoma cirratum*; nurse shark), and martillo (*Sphyrna mokarran*; great hammerhead shark), although these species have the potential to be integrated into ecotourism activities (Cisneros-Montemayor et al., 2020). This would be an example of how the SES methodology could be used to determine the kinds of activities that contribute to the economy while also integrating fishers in the decision-making and use of this resource.

The potential options to improve the shark fishery in this location are: 1) issue regional fisheries management that integrates local knowledge, 2) update the seasonal closure for both size categories, 3) implement a period of closure for rays, and/or 4) generate added value and optimal use of resources (for example, shark skin that is used in households as a cleaning material). Consequently, assigning individual species to catch quotas for fishers with commercial permissions would maximize their economic gains in a sustainable manner. In addition to strengthening dockside inspection, surveillance, and regulation enforcement, authorities must ensure that fishers with larger vessel scale permits remain accountable for meeting their catch quotas and generate dialogue to strengthen co-responsibility in the management of elasmobranch species.

FISHER VOICES AND THEIR ROLE IN THE ADVANCEMENT OF LOCAL ECOLOGICAL KNOWLEDGE

It is estimated that more than 10,000 tonnes of sharks are landed every year in Nigeria through target fishing and also as bycatch (FDF, 2015). A mean catch of 19,008 MT was recorded between 2007 and 2017 making her one of the top ten catchers during this period (FAO FishStat, 2019). Nigeria was also reported to land the largest catch in Africa (over 32,000 tonnes) in 1983 (Compagno, 1994). Despite the large number of catches there has not been much scientific work done to understand the status of shark fisheries in Nigeria. Understanding the trends of a nation's fisheries can help scientists to observe changes in species' abundance, predict species' distribution, discover new species, and detect those close to extinction. This information can assist fisheries' managers in making appropriate decisions for management.

Shark scientists and conservationists have often resorted to knowledge from past and present shark fishers, local resource users, and coastal dwellers to determine past and present trends in Nigerian shark fisheries. This is known as *local ecological knowledge (LEK)*. LEK is the knowledge of local ecosystems obtained by a community through years of keen observation. Scientists have been using this tool in recent years to improve the state of elasmobranch conservation.

Minority scientists often find it easier to gain access to the local folks which can assist with closing the gap that may arise from socioeconomic, cultural, and even political disconnect which often happens between the local resource users and scientists from large institutions who aren't from the community. Information obtained from LEK has been really helpful for scientists in determining the status of sharks in Nigeria. Local shark fishers, especially the old and retired ones, love to reminisce on years when they were active fishers and once they are comfortable with the presence of a scientist (which can happen more quickly if they share

143

an identity or familiarity with that scientist) they eagerly share their knowledge (Figures 4.6 and 4.7).

While LEK can provide interesting discussions and new information, the responsibility of confirming these deductions lies with the research scientists who collect, analyze, and interpret the data. Anecdotal evidence suggests that the reason local fishers find it difficult to share their knowledge is because they perceive scientists as threats lying in wait to prosecute them for catching sharks (Figure 4.8).

It is important to let them know that the purpose of collecting this information is to assist elasmobranch conservation. Several species of sharks are endangered and should not be caught while other species should be managed to avoid further depletion. With proper management practices in place, future generations of fishers would enjoy good catch rates of sharks. With these explanations fishers often feel more relaxed

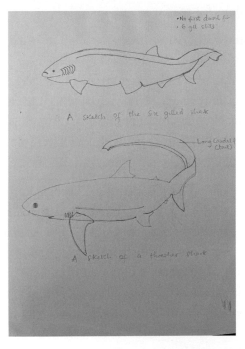

Figure 4.6 *Illustrations of the six gilled shark (top) and thresher shark (bottom) based on the old fishers description.*

Figure 4.7 Picture of a landed thresher shark.

Figure 4.8 Picture of Lara Fola-Matthews and a long-time shark fisher friend.

and have a better understanding that scientists and fisheries' officials are only working towards sustainability of the species.

From what can be deduced from the LEK gathered by interacting with Nigerian fishers, studies on the status, biology, and ecology of various shark species in Nigeria is necessary. It would greatly assist in making proper management plans for species and also improve ongoing efforts in developing a National Plan of Action (NPOA) of sharks for Nigeria.

Figure 4.9 *Picture of Lara Fola-Matthews and pupils at a school during an engagement.*

The NPOA-Sharks are guidelines specific to a nation for the conservation of its shark fisheries. It should provide information on the regulatory framework, management, structure, monitoring, and enforcement of laws related to the trade of sharks and shark by-products (Kerwath et al., 2013) (Figure 4.9).

THE PRESENT

UNDERSTANDING SMALL-SCALE FISHERIES THROUGH INTERVIEWS

Although research and conservation efforts have mostly focused on the effects of industrial fisheries, small-scale fisheries account for more than 95% of fishers in the world, especially in countries of the Americas, Africa, and the Indo-Pacific region (Pauly, 2006). Given their wide occurrence and the large number of dependents, small-scale fisheries are an important economic sector (Johnson et al., 2013) and their impact on elasmobranchs may be significant (Caceres et al., 2022; Hawkins & Roberts, 2004; Salas et al., 2007; Moore et al., 2010).

Elasmobranchs caught through artisanal fisheries may be targeted (meaning that elasmobranchs are highly preferred and purposely landed e.g. Cartamil et al., 2011) or caught as bycatch (meaning that they aren't targeting elasmobranchs but tend to catch some e.g. Baeta et al., 2010). Either way, the catch and landing data for elasmobranchs in artisanal fisheries unfortunately tends to be rather spotty, making it difficult to examine trends over time. Some factors which contribute to this problem are poor monitoring and regulation of *subsistence fishing*, recording practices which disregard bycatch and/or species perceived to be of lesser economic value, and the fact that landing sites may be remote and hard for data collectors to access (Pauly et al., 2002). This leads to problems like failure to identify elasmobranchs to species level and underreporting catch data to governing agencies (Baeta et al., 2010), which makes such fisheries difficult to understand and manage.

USING THE INTERVIEW METHOD WITH SMALL-SCALE FISHERS

Interview surveys can enhance understanding of the interactions between artisanal fisheries and marine taxa, particularly charismatic and recognizable species such as marine mammals (dolphins and whales), elasmobranchs, and sea turtles (Hall & Close, 2007; Moore et al., 2010, Hind et al., 2014). Despite the limitations of social survey data (e.g. data are generally more qualitative than quantitative), the interview method provides insights into species targeted and caught, quantities captured, and gear types used in a low-cost and time-efficient manner (Moore et al., 2010; Carruthers & Neis, 2011; Tesfamichael et al., 2014). Using fishers' knowledge can also clarify current and historical catch information and can help integrate stakeholders in research and conservation efforts. Support from stakeholders is an important part of natural resource management; interviews are an excellent opportunity for the researcher to build rapport with the relevant communities and it's a great way to incorporate stakeholder input into conservation.

CONDUCTING SURVEYS AND INTERVIEWS

There are a variety of interview and sampling methods to collect interview data from fishers. For example, key informant interviews are in-depth, semi-structured conversations with highly knowledgeable community members, and an excellent way to gain detailed qualitative insights into the fishery. However, to collect a mix of quantitative and qualitative data, most interview surveys tend to follow the *knowledge, attitudes and practices (KAP)* framework.

KNOWLEDGE, ATTITUDES AND PRACTICES SURVEYS

KAP surveys were first developed in the 1950s and have since become standard practice in many fields, including in medical, behavioral, and health education research. The KAP method is cost-effective and conserves resources more than other social research methods, since questions are focused and limited in scope (Eckman et al., 2008). A KAP survey has predefined questions in a standardized format that provides quantitative and qualitative information on the topic of interest, such as shark catches in a certain fishery. KAP surveys can also reveal misconceptions that may represent obstacles to conservation and/or management.

KAP surveys generally contain four sections: Demographics, Knowledge Questions, Attitude Questions, and Practice Questions. During analysis, the knowledge, attitudes, and practices of various demographics can be explored to look for interesting relationships. For example, a KAP survey could answer the question "Is there a relationship between the age of the fishers and how many shark species they recognize?" by collecting data on the age of fishers in the Demographics section and asking about which shark species they recognize in the Knowledge section, then using that data in analyses.

Some common question types used in KAP surveys to gather quantitative data include binary (i.e. yes/no), multiple response (more than one option), and Likert Scale questions, which are used to measure agreement with various attitudes via a 5 or 7 point scale (e.g. strongly disagree, disagree, neutral, agree, strongly agree). Qualitative data can be collected using open-ended questions. Other important data include response rate and rejection rate, which indicates how well your data collection effort went and if it adequately represents the sample population.

Once survey questions have been drafted under a KAP framework, some common ways to conduct interviews and surveys are via internet, telephone, or in person. Researchers should choose methods which will give them the greatest access to their target population. The following examples illustrate appropriate selection of survey methods.

A study done in Florida analyzed websites and collected online surveys of charter boat captains to assess the scale of the state's charter boat shark fishing industry. Online surveys are relatively easy and cost-effective to execute, but results may be biased to exclude demographics with decreased access to technology. Most of the fishing charter companies in the study area used websites to book clients, indicating that the target population was heavily reliant on internet access and that using online surveys was an appropriate approach. Using the KAP framework, researchers determined that recreational charter boat shark fishing occurs throughout Florida, that shark fishing is often the most expensive trip offered, and that fishers that show a strong conservation ethic toward sharks commonly practice catch and release (Shiffman & Hammerschlag, 2014).

A study done in Trinidad and Tobago (where shark meat is a popular culinary delicacy) interviewed citizens in person about their knowledge and consumption of sharks. In-person interviews are limited by the times and places where data can be collected, but they capture a wide range of participants since they are not reliant on the target population's access to or comfort with technology. The aim of the study was to reach a wide public that may have participated in shark consumption; therefore, doing in-person interviews at community hubs was the best method to gather information on potential shark consumers. This study also followed the KAP framework and found that most participants displayed attitudes in favor of shark conservation and sustainable use, that over 70% of respondents ate shark, and over half ate shark infrequently enough to avoid risks from heavy metal toxicity (Ali et al., 2020).

Similarly, a study done in the Caribbean Coast of Colombia interviewed fishers about their fishing practices and shark knowledge using the KAP framework.

The aim was to capture fishing practices and shark fishing perceptions, so in-person interviews were conducted at fishing docks, fish cleaning stations, and fishers' homes. In order to gather information from fishers, one must be considerate to not interrupt their work schedule (and therefore potential income) while also approaching them at a time when their thoughts are engrossed with fishing and most receptive to talk about this subject.

The study concluded that although fishers do not target sharks, most fishers will keep shark catches for consumption or to sell, that the majority fishers perceived decline of sharks since they started fishing, and that effective community co-management and stakeholder inclusion are needed to ensure effective conservation strategies (Caceres et al., 2022).

These examples show how important it is to consider the context of data collection, including the format, location and timing, when interpreting the results of a report or study.

Stakeholders (scientists, fishers, and the public alike) should always be aware of the limitations and benefits of different data collection methods.

IMPLICATIONS FOR FISHERIES MANAGEMENT: THE IMPORTANCE OF UNDERSTANDING SMALL-SCALE FISHERIES

The long-term sustainability of many marine ecosystems is threatened. Conventional efforts to manage these systems have proven to be insufficient, whether species-specific or per fishing sector (Mascia, 2003). Nonetheless, government entities around the world have taken measures to manage marine resources, such as creating Marine Protected Areas, sanctuaries, issuing fishing permits, and instituting fishing regulations.

Countries that agreed to the *FAO*'s International Plan of Action (IPOA) for the management of shark resources are expected to create and implement their own National Plan of Action (NPOA).

Historically, there has been a pessimistic outlook on the possibility that stakeholders will voluntarily cooperate, which has led to a widespread use of a centralized power to manage common resources. However, oftentimes the centralized power is so removed from the community it manages, that the creation and enforcement of rules can lead to resentment and resistance from the community. Further research is needed to illustrate the social, economic, and cultural drivers behind the demand for shark catches in small-scale fisheries in order to understand the full extent of reliance on elasmobranchs, particularly among underrepresented communities. This includes what proportion of protein intake is derived from elasmobranchs. Future research should include interview surveys as a method to complement established field data collection, and future studies should have an inherent component of capacity building and stakeholder inclusion in order to improve transparency in data collection and increase support from local communities for conservation policies.

ELASMOBRANCH ECOLOGY IN BRAZIL: A CASE STUDY IN THE STATE OF SANTA CATARINA

Brazil is one of the largest countries in the world and is rich in biodiversity of all forms of life (MMA, 2021). Of the more than 1400 species of cartilaginous fish (fish with skeletons made of cartilage) worldwide, 89 species of sharks, and 70 species of rays live in Brazilian waters (Fricke et al., 2021; ICMBIO, 2016). These species range from shallow coastal waters to deep-sea regions and some stingray species (Family Potamotrygonidae) are even adapted to live in freshwater habitats (Charvet-Almeida et al., 2005; Rosa & Gadig, 2014). The scientific study of elasmobranchs in Brazil began with the arrival of foreign scientific expeditions in 1872 (Lessa et al., 1999). During the decades of the 1960s–1970s, some research was conducted by Sadowsky (1965) on coastal sharks and rays that were landed in the region of São

Paulo, but other coastal and oceanic regions were more difficult to access and remained poorly studied. In the 1970s–1980s, according to Lessa et al. (1999), more scientific cruises operated in Brazilian waters which resulted in the description of new species of endemic deep-sea skates, such as one fin skate (*Gurgesiella dorsalifera*; McEachran & Compagno, 1980), thintail skate (*Dipturus leptocauda*; Krefft & Stehmann, 1975) and whitemouth skate (*Bathyraja schroederi*; Krefft, 1968). In the 1990s, a program named REVIZEE (Assessment Program for the Sustainable Potential of Living Resources in the Exclusive Economic Zone) was created with the objective to collect and identify marine organisms throughout the entire Brazilian coast and more species were found (MMA, 2006, Rincon et al., 2017). Carneiro (2007) noted that this program was an important milestone for the study of several marine species, including elasmobranch.

Since then, the number of elasmobranch studies has increased in Brazil (Coelho et al., 2021); however, there was no opportunity to conduct a baseline study of the overall populations in Brazil because many species were overfished and disappeared (Vooren & Klippel, 2005). Due to the growth in human population in Brazil and worldwide, there has been an increase in pressure for harvesting coastal and oceanic resources (Wever et al., 2012; Abdallah & Sumaila, 2007). Despite the existent federal fishery laws, a few national and international commercial fishing boats continue to work illegally in areas and seasons that are permanently or temporarily closed to fishing in Brazil (Brandini, 2014). This behavior poses a threat to several species that are already endangered or that could become endangered, since the fishing reports are not always committed to the truth and captures are underreported or not reported at all (Barreto et al., 2017; Sumaila et al., 2020).

In South Brazil, there is a city called Balneário Camboriú, located in the Santa Catarina state and it is known for having the tallest buildings in Latin America, which makes people call it "Brazilian Dubai." The average population is 150,000 inhabitants for most of the year, but during the summer season, especially New Year's, it receives more than 1 million people in the Central beach (Codesso et al., 2013; IBGE, 2021).

The economy is now based on tourism and civil construction, but a few decades ago it was an artisanal fishing community, and some fishers are still operating. Back in the 1920s the city was home for a huge artisanal fishing community (Schlickmann, 2016). One of the most traditional forms of fishing is beach seining which targets fish living close to the shore line (Da Silva, 2002). It consists of a canoe manually pushed through the sand to the water, then fishers release a large enclosing net in a circular shape then drawn ashore by hand from both ends (Gamba, 1994). It has a strong influence on the culture and mores of the state, especially the fishing tradition involved in lebranche mullets (*Mugil liza*) (Bannwart, 2013; De Souza et al., 2017). The rest of the year, fishers target other species, such as White mullet (*Mugil curema*), snook (*Centropomus* spp.), Whitemouth cracker (*Micropogonias furnieri*), and king croaker (*Menticirrhus* spp.).

Given several reports, photos, and videos of sharks and rays being caught incidentally by a beach seine, a few attempts were made to monitor the fishery in 2018; however, there was resistance and mistrust from the fishers towards researchers because they were afraid of being harmed somehow because of the information that would be collected and published. In May 2020, during the COVID-19 pandemic, fishers decided to cooperate with a monitoring program. Since the beginning of the fishery, cownose stingrays (*Rhinoptera bonasus* and *R. brasiliensis*) have been caught alive and due to their high presence in beach seines, a specific program was created to study them. At first, only biometric data was collected, such as total length, disk width, and total weight. During the winter the stingrays started to show ectoparasites, becoming less responsive and a few even died. After researchers observed these things, they started collecting blood samples of cownose rays with and without ectoparasites to analyze what was happening to their health and how parasites were affecting them. The preliminary results showed that cownose stingrays have their blood count altered by the presence of ectoparasites.

In addition to cownose rays, other elasmobranch species were caught by a beach seine in the Central Beach of Balneário Camboriú, such as Spiny butterfly ray (*Gymnura altavela*), longnose stingray (*Hypanus guttatus*), Brazilian sharpnose shark (*Rhizoprionodon lalandii*), and scalloped hammerhead shark (*Sphyrna lewini*). There are reports of trawling fishers accidentally capturing Rio skate (*Rioraja agassizii*), lesser guitarfish (*Zapteryx brevirostris*), and old pictures of fishers with sand tiger sharks (*Carcharias taurus*), but *C. taurus* has since been fished to the point where sightings are extremely rare. The Central Beach of Balneário Camboriú is hypothesized to be an essential habitat for a part of the life cycle of several threatened elasmobranch species. More research needs to be conducted to confirm this hypothesis but unfortunately there are limitations due to the lack of funding for the project, and so the feeding, reproductive, and other areas of study are still unknown.

At the beginning of March 2021, there was a forced stoppage in fishing activities and consequently in research due to the start of the beach nourishment work. The food security and income of artisanal fishers in Central Beach were affected until the completion of the work, which took place in October 2021. The possible impacts caused by the beach nourishment on the sharks and rays that inhabited this region also remain unstudied. However, fishers have already noticed a reduction in the number of individuals of several fish species that are caught by beach seine when compared to previous years; so in order to evaluate what is happening there, a new and long-term monitoring program should be done.

In addition to the scientific research, an education and awareness program was carried out with fishers with the goal of educating them and exchanging knowledge about the target and non-target species that are caught by them. The biggest result is through collaboration with fishers; from May 2020 to December 2021, more than 750 cownose stingrays have been released after being accidentally caught by a beach seine. Since the research is conducted on the beach in a public setting, it serves as a unique opportunity for the community and tourists to see and learn about the species. People are frequently asking to participate during the process of collecting data, so they help contain and weigh the stingrays. By far the most interested audience is children, who love to watch and ask questions. Sometimes, kids leave saying to their parents that they want to be a marine biologist. Through community science and fishers collaboration there is an opportunity to develop research on coastal sharks and rays in Brazil and create participatory management and conservation plans (Figures 4.10–4.13).

FISHERIES-DEPENDENT STUDIES IN THE PERSIAN GULF

The Persian Gulf is a sea in Western Asia. The body of water is an extension of the Arabian Sea (Gulf of Oman) through the Strait of Hormuz and lies between Iran to the northeast and the Arabian Peninsula to the southwest. The Persian Gulf has many fishing grounds, extensive reefs (mostly rocky, but also coral), and abundant pearl oysters, but its ecology has been damaged by industrialization and oil spills (Emery, 1956). Overall, the wildlife of the Persian Gulf is endangered from global factors as well as regional and local negligence. Most

Figure 4.10 Researcher collecting a blood sample with the help of an eight-year-old girl named Eloa, sister of the youngest fisherman.

Figure 4.11 *Fisher helping in the process of collecting biometric data from a longnose stingray* (Hypanus guttatus) *incidentally caught by a beach seine.*

Figure 4.12 *Researcher releasing a cownose ray* (Rhinoptera bonasus) *at Central Beach in Balneário Camboriú, South Brazil.*

Figure 4.13 *Fisherman releasing a cownose ray (*Rhinoptera bonasus*) at Central Beach in Balneário Camboriú, South Brazil. Photo credit: Noemia Venturi.*

pollution is from ships, but land-generated pollution is the second most common source of pollution (Khazali, 2021). The Persian Gulf is home to over 700 species of fish, most of which are native. Of these 700 species, more than 80% are reef-associated (Eagderi et al., 2019). Recent years have seen a drastic decline in the coral population in the Persian Gulf, partially owing to climate change, but mostly due to irresponsible dumping of construction garbage such as tires, cement, and chemical by-products by other countries which have found their way to the Persian Gulf in recent years (Khazali,2021). Aside from direct damage to the coral, the construction waste creates "traps" for marine life in which they are trapped and die. The result has been a dwindling population of the coral, and as a result a decrease in the number of species that rely on the corals for their survival (Shepperd, 2016). Due to its semi-enclosed nature, it has deep waters as well as extensive reefs. Its connection to the Indian Ocean also makes it rich with migratory and non-migratory habitat of almost 30 species of sharks (Jabado et al., 2015). Unfortunately, in this area, just a few studies have been done to identify the sharks and most of these studies mostly rely on fisher information and their bycatch. There are few genetic or molecular studies about sharks and rays in this area, especially in Iran. In addition, no systematic program has been devoted to shark conservation.

In an effort to begin identifying sharks and rays in this area and studying their behavior, Raysha Persia, an NGO founded by Sara Asadi, started its shark research program. The first method Raysha Persia implemented was the usage of the BRUV (baited remote underwater video) system in shallow water and deep water in Kish Island in Iranian waters. This system is used in marine biology research by attracting fish into the field of view of a remotely controlled camera. The technique records fish diversity, abundance, and behavior of species. Sites are sampled using a video camera on the bottom of the sea which records the region surrounding a baited canister. Baited cameras are highly effective at attracting predators and are a non-invasive method of generating relative abundance indices for a number of marine species. The captured images and videos can be analyzed and used for other studies (Kilfoil et al., 2021).

The exploration study was run for over a year and black tip reef (*Carcharhinus melanopterus*), white tip reef (*Triaenodon obesus*), and scalloped hammerhead shark (*Sphyrna lewini)* were identified as the most common species in that area. In mid-2018, the research team started using a BRUV system in Hendurabi and in 2019 they began using this system on the Halul Island of Qatar to understand the relationships between the sharks living in different regions of the gulf. In the second stage of this project, the mangrove habitat situated in Qeshm Island, which is a protected nursery area for sharks, was explored to gain a better understanding of the abundance of baby sharks. Furthermore, the team also trained the local community and fishers about the importance of sharks and gave them questionnaires and booklets with the pictures of species so they could mark anytime that they saw sharks and later they reported their observations to the researcher.

UNDERSTANDING THE CATCH DATA OF ELASMOBRANCHS IN DATA-DEFICIENT REGIONS AND THE ROLE OF A CONSERVATIONIST

Sri Lanka is surrounded by many habitat types ranging from mangrove forests, seagrass beds, coral reefs, and lagoons (Dayananda, 1992), Elasmobranchs (sharks and rays). According to fishers in Sri Lanka, elasmobranchs could be found in these habitats and many species may move between several habitats in the waters surrounding this island nation. Highly migratory species tend to spend most of their time in open waters and come to shallower waters for breeding. The coastline around Sri Lanka could be essential habitat for elasmobranchs, and it is believed to have breeding grounds for many species of elasmobranchs (Buddhi personal communication, March 2019).

It is revealed that updated elasmobranch checklists featured nearly 110 to 120 species in Sri Lanka (De Silva, 2006, 2015; Last et al., 2016b; Ebert et al., 2017; Fernando et al., 2019) However, drastic population declines have been reported from the commercially harvested elasmobranchs in Sri Lanka (SLNPOA-Sharks, 2013; Davidson et al., 2016; Fernando et al., 2019). This is the biggest challenge faced by conservationists, marine biologists, and policy-makers in Sri Lanka. This is why understanding elasmobranch status and their role in the waters of Sri Lanka is so important.

The primary objective of Blue Resources Trust (BRT) is to develop a long-term monitoring program for elasmobranchs caught in the *EEZ* including high seas and coastal waters around Sri Lanka. This includes the development of data collection and best-practice toolkits to influence the data collection of similar fisheries around Sri Lanka, and to execute and continue capacity-building workshops for fisheries managers, customs officers, and other stakeholders to have long-lasting sustainable fisheries in Sri Lanka.

In August 2017, BRT expanded their mobulid (manta and devil ray) project to identify all elasmobranch diversity in Sri Lanka. Since then, the organization has conducted more than 1,000 surveys across the country covering the north, east, south, and west coasts. Up to now, BRT has recorded more than 26,900 specimens comprising 53 shark and 50 ray species. Out of these 103 species, 12 species are the first records of these animals in Sri Lanka (BRT, unpublished data) and 4 species are potentially undescribed anywhere in the world (Fernando & Stewart, 2021; Fernando et al., 2019).

Shark species, especially silky and blue sharks are heavily exploited by the larger single- and multi-day vessels operating within the EEZ and in high seas (Figure 4.14). The coastal shark species like milk sharks and grey sharpnose sharks are harvested by small coastal fishing boats (Outboard Fiber Reinforced Plastic Boats—OFRP) with outboard engines and also by small canoes. Shark and ray meat is consumed locally (Figure 4.15), and shark fins are exported to countries like China, Hong Kong, and Singapore as there is a high demand for shark fin soup in these countries. There is no escape even for bottom-dwelling shark species as they are targeted to obtain shark liver oil. There is no demand for deep-sea shark meat but in several landing sites, their meat is salted, sun-dried, and sold to local consumers. Shark liver oil is exported to Japan in raw form to extract squalene, which is used in the cosmetics and medical sectors.

Figure 4.14 *Pile of finned shark at one of the major harbors in Sri Lanka.*

Figure 4.15 *Drying* Aetobatus ocellatus *(white spotted eagle ray) in the northwest coast of Sri Lanka.*

Figure 4.16 Rhynchobatus australiae *(white spotted guitarfish) is harvested by south OFRP boats in the bottom gill net.*

On the other hand, rhino rays are captured as bycatch and landed as they have highly valued fins compared to other elasmobranch species. The coastal waters around Sri Lanka are home to many rhino rays and up to now, six species of rhino rays have been recorded including two wedgefishes (*Rhina ancylostoma* and *Rhynchobatus australiae* (Figure 4.16)), three guitarfishes (*Acroteriobatus variegatus*, *Rhinobatos annandalei*, and *Rhinobatos lionotus*), and one giant guitarfish

(*Glaucostegus granulatus*). Out of these, four species (*R. ancylostoma, R. australiae, A. variegatus, R. lionotus* and *G. granulatus*), are critically endangered and one species (*R. annandalei*) is data-deficient according to the International Union for Conservation of Nature Red List of Threatened Species (IUCN Red List, 2021). Many specimens of giant guitarfish (*G. granulatus*) are being landed by the small coastal fishing (OFRP) boats along the northern coast of Sri Lanka (Figure 4.17).

With over a decade of combined experience, BRT has fine-tuned the data collection methods for artisanal and small-scale fisheries and improved skills to distinguish similar species in the same genus. BRT also supports the implementation of national and international regulations through training and capacity building of several national government authorities on proper identification of shark fins and the development of national policies, such as the National Plan of Action for Sharks. BRT also conducts awareness programs for school children, university students, and children in fisher villages. BRT strongly believes that it is not sufficient to stop at the publication level but that it is essential to influence authorities to protect elasmobranchs that have extremely depleted populations with slow chances of recovery, in addition to supporting transitions into sustainable fisheries. For instance, a recent paper on the collapse of the sawfish populations in Sri Lanka was published by BRT

Figure 4.17 *Pile of juvenile cownose from OFRP boat.*

(Tanna et al., 2021) and made available to national policy-makers, who are now in the process of identifying suitable management measures to effectively protect these species.

CONCLUSION

Fisheries management around the world looks different depending on the socioeconomic dynamics of different regions. This chapter discussed the variety of ways data-deficient fisheries are managed in countries that rely on elasmobranch species for a community's livelihood, food security, and a lasting cultural history. While conventional methods for modeling fishery stocks and determining the health of stocks are readily available, data-deficient fisheries require a more diverse pool of data that is not limited to solely quantitative data. In this chapter, a variety of diverse and unconventional research techniques are explained in the hopes of encouraging the continued recognition of sociological influences in fisheries science. Through interviews, historic information gathering, evaluation of ecosystem services, and integration of local knowledge, shark populations can be further assessed with the best available data. Fisheries science is multifaceted, meaning it does not only include biology, but it also includes economics and a basic understanding of sociological principles. Identifying important species that may be more vulnerable to fishing pressures coupled with the basic understanding of their biology has allowed for the development of methodologies that cater to the different countries that rely on small-scale fisheries rather than implementing blanket regulations based on incomplete or insufficient data.

REFERENCES

Abdallah, P. R., & Rashid Sumaila, U. (2007). An historical account of Brazilian public policy on fisheries subsidies. *Marine Policy, 31*(4), 444–450. https://doi.org/10.1016/j.marpol.2007.01.002.

Ali, L., Grey, E., Singh, D., Mohammed, A., Tripathi, V., Gobin, J., & Ramnarine, I. (2020). An evaluation of the public's knowledge, attitudes and practices (KAP) in Trinidad and Tobago regarding sharks and shark consumption. *PLOS ONE*, *15*(6), e0234499. https://doi.org/10.1371/journal .pone.0234499.

Almerón-Souza, F., Sperb, C., Castilho, C. L., Figueiredo, P. I. C. C., Gonçalves, L. T., Machado, R., … Fagundes, N. J. R. (2018). Molecular identification of shark meat from local markets in Southern Brazil based on DNA barcoding: Evidence for mislabeling and trade of endangered species. *Frontiers in Genetics*, *9*(April), 1–12. https://doi.org/10.3389/fgene.2018.00138.

Baeta, F., Batista, M., Maia, A., Costa, M. J., & Cabral, H. (2010). Elasmobranch bycatch in a trammel net fishery in the Portuguese West Coast. *Fisheries Research*, *102*(1–2), 123–129. https://doi.org/10.1016/j.fishres.2009.10.016.

Baker-Médard, M., & Faber, J. (2020). Fins and (mis)fortunes: Managing shark populations for sustainability and food sovereignty. *Marine Policy*, *113*(February 2019). https://doi.org/10.1016/j.marpol.2019.103805.

Bannwart, J. P. (2013). A pesca da tainha no litoral catarinense. *Agropecuária Catarinense*, *26*(2), 15–18.

Barreto, R. R., Bornatowski, H., Motta, F. S., Santander-Neto, J., Vianna, G. M. S., & Lessa, R. (2017). Rethinking use and trade of pelagic sharks from Brazil. *Marine Policy*, *85*, 114–122. https://doi.org/10.1016/j.marpol.2017.08.016.

Baum, J. K., Myers, R. A., Kehler, D. G., Worm, B., Harley, S. J., & Doherty, P. A. (2003). Collapse and conservation of shark populations in the Northwest Atlantic. *Science*, *299*(5605), 389–392.

Bonfil, R. (1997). Status of shark resources in the Southern Gulf of Mexico and Caribbean: Implications for management. *Fisheries Research*, *29*(2), 101–117.

Bornatowski, H., Braga, R. R., Kalinowski, C., & Vitule, J. R. S. (2015). 'Buying a Pig in a Poke': The problem of elasmobranch meat consumption in Southern Brazil. *Ethnobiology Letters*, *6*(1), 196–202. https://doi.org/10.14237/ebl.6.1 .2015.451.

Braccini, M., Blay, N., Harry, A., & Newman, S. J. (2020). Would ending shark meat consumption in Australia contribute to the conservation of white sharks in South Africa? *Marine Policy*, *120*(July), 104144. https://doi.org/10 .1016/j.marpol.2020.104144.

Brandini, F. (2014). Marine biodiversity and sustainability of fishing resources in Brazil: A case study of the coast of Paraná State. *Regional Environmental Change*, *14*(6), 2127–2137. https://doi.org/10.1007/s10113-013-0458-y.

Bundy, A. (2005). Structure and functioning of the Eastern Scotian Shelf ecosystem before and after the collapse of groundfish stocks in the early 1990s. *Canadian Journal of Fisheries and Aquatic Sciences, 62*(7), 1453–1473.

Caceres, C., Kiszka, J. J., Luna-Acosta, A., Herrera, H., Zarza, E., & Heithaus, M. R. (2022). Predatory fish exploitation and relative abundance in a data-poor region from the Caribbean coast of Colombia, inferred from artisanal fisheries interview surveys and Baited remote underwater videos. *Aquatic Conservation: Marine and Freshwater Ecosystems.*

Cailliet, G. M. (2015). Perspectives on elasmobranch life-history studies: A focus on age validation and relevance to fishery management. *Journal of Fish Biology, 87*(6), 1271–1292.

Calich, H., Estevanez, M., & Hammerschlag, N. (2018). Overlap between highly suitable habitats and longline gear management areas reveals vulnerable and protected regions for highly migratory sharks. *Marine Ecology Progress Series, 602*, 183–195.

Carneiro, M. H. (2007). Diagnóstico dos recursos pesqueiros marinhos, Cynoscion jamaicensis, Macrodon ancylodon e Micropogonias furnieri (perciformes: sciaenidae), da região sudeste-sul do Brasil entre as latitudes 23° e 28° 40 S.

Carruthers, E. H., & Neis, B. (2011). Bycatch mitigation in context: Using qualitative interview data to improve assessment and mitigation in a data-rich fishery. *Biological Conservation, 144*(9), 2289–2299. https://doi.org/10.1016/j.biocon.2011.06.007.

Cartamil, D., Santana-Morales, O., Escobedo-Olvera, M., Kacev, D., Castillo-Geniz, L., Graham, J. B., … Sosa-Nishizaki, O. (2011). The artisanal elasmobranch fishery of the pacific coast of Baja California, Mexico. *Fisheries Research, 108*(2–3), 393–403. https://doi.org/10.1016/j.fishres.2011.01.020.

Castillo-Géniz, J. L., Márquez-Farias, J. F., Cruz, M. C. Rdl, Cortés, E., & Prado, A. Cd (1998). The Mexican artisanal shark fishery in the Gulf of Mexico: Towards a regulated fishery. *Marine and Freshwater Research, 49*(7), 611–620.

Charvet-Almeida, P., de Araújo, M. L. G., & de Almeida, M. P. (2005). Reproductive aspects of freshwater stingrays (Chondrichthyes: Potamotrygonidae) in the Brazilian Amazon Basin. *Journal of Northwest Atlantic Fishery Science, 35*, 165–171. https://doi.org/10.2960/j.v35.m502.

Cisneros-Montemayor, A. M., Becerril-García, E. E., Berdeja-Zavala, O., & Ayala-Bocos, A. (2020). Shark ecotourism in Mexico: Scientific research, conservation, and contribution to a blue economy. *Advances in Marine Biology, 85*(1 January), 71–92. https://doi.org/10.1016/bs.amb.2019.08.003.

Codesso, M. M., Lunkes, R. J., & Suave, R. (2013). Práticas Orçamentárias Aplicadas Em Empresas Hoteleiras No Brasil: Um Estudo Na Cidade De Balneário Camboriú – SC. *Turismo - Visão e Ação, 15*(2), 279. https://doi.org/10.14210/rtva.v15n2.p279-294.

Coelho, K. K., Lima, F. S., Wosnick, N., Nunes, A. R. O. P., Silva, A. P. C., Gava, T. T., ... Nunes, J. L. S. (2021). Research trends on elasmobranchs from the Brazilian Amazon coast: A four-decade review. *Biota Neotropica, 21*(4). https://doi.org/10.1590/1676-0611-bn-2021-1218.

Compagno, L. J. V. (1994). *Preliminary report for the sub-equatorial African region, Atlantic, Indian, and Antarctic Oceans.* International Union for the Conservation of Nature, Shark Specialist Group, 50 pp.

Cortés, E. (1999). Standardized diet compositions and trophic levels of sharks. *ICES Journal of Marine Science, 56*(5), 707–717.

Da Silva, P. (2002). *Common property to co-management: Social change and participation in Brazil's first maritime extractive reserve.* London School of Economics and Political Science. PhD dissertation. http://etheses.lse.ac.uk/id/eprint/2287.

Davidson, L. N. K., Krawchuk, M. A., & Dulvy, N. K. (2016). Why have global shark and ray landings declined: Improved management or overfishing? *Fish and Fisheries, 17*(2), 438–458.

Dayananda, H. V. (1992). *Shoreline erosion in Sri Lanka's coastal areas.* Colombo: Coast Conservation Department.

De Silva, R. I. (2006). Taxonomy and status of the sharks and rays of Sri Lanka. In C. N. Bambaradeniya (Ed.), *Fauna of Sri Lanka: Status of taxonomy, research and conservation* (pp. 294–301). Colombo, Sri Lanka: The World Conservation Union (IUCN).

De Silva, R. I. (2015). *The sharks of Sri Lanka.* Colombo: Field Ornithology Group of Sri Lanka.

De Souza, D. S., Pereira Silva, R. C., & Steenbock, W. (2017). De Quem É o Peixe? Aspectos Socioeconômicos Da Pesca Industrial e Artesanal De Tainha (Mugil Liza) Em Santa Catarina. *Revista CEPSUL - Biodiversidade e Conservação Marinha, 6.* https://doi.org/10.37002/revistacepsul.vol6.665e2017002.

del Pilar Blanco-Parra, M., AlbertoNiño-Torres, C., Ramírez-González, A., & Sosa-Cordero, E. (2016). Tendencia Histórica de La Pesquería de Elasmobranquios En El Estado de Quintana Roo, México. *Ciencia Pesquera*, Número especial *24*, 125–137.

Dent, F., & Clarke, S (2015). State of the global market for shark products. FAO Fisheries and Aquaculture technical Paper 590, I.

DOF. de la Federación, Diario Oficial. (2007). *Ley general de pesca y acuacultura sustentables.* Ciudad de México, México: Gobierno Federal Mexicano.

Dulvy, N. K., Baum, J. K., Clarke, S., Compagno, L. J. V., Cortés, E., Domingo, A., ... Valenti, S. (2008). You can swim but you can't hide: The global status and conservation of oceanic pelagic sharks and rays. *Aquatic Conservation: Marine and Freshwater Ecosystems, 18*(5), 459–482.

Dulvy, N. K., Fowler, S. L., Musick, J. A., Cavanagh, R. D., Kyne, P. M., Harrison, L. R., ... White, W. T. (2014). Extinction risk and conservation of the world's sharks and rays. *eLife, 3*, e00590.

Eagderi, S. et al. (2019). Annotated checklist of the fishes of the Persian Gulf: Diversity and conservation status. *Iranian Journal of Ichthyology, 6*, 1–171.

Ebert, D. A., De Silva, R. I., & Goonewardena, M. L. (2017). First record of the dwarf false catshark, *Planonasus parini (Carcharhiniformes: Pseudotriakidae)* from Sri Lanka. *Loris, 27*, 63–64.

Eckman, K. & Walker, R., with Bouapao, L., & Nuckles, K. (2008). *Non-point source pollution (NPS) project evaluation practices in Minnesota: Summary report.* Saint Paul, MN: Water Resources Center, University of Minnesota.

Emery, K. O. (1956). Sediments and water of Persian Gulf. *AAPG Bulletin, 40*(10), 2354–2383.

Espinoza-Tenorio, A., Wolff, M., Taylor, M. H., & Espejel, I. (2012). What model suits Ecosystem-based fisheries management? A plea for a structured modeling process. *Reviews in Fish Biology and Fisheries* 22 (1), 81–94.

Essington, T. E., Beaudreau, A. H., & Wiedenmann, J. (2006). Fishing through marine food webs. *Proceedings of the National Academy of Sciences, 103*(9), 3171–3175.

FAO. (2019). Fishery and aquaculture statistics. Global capture production 1950–2017 (FishstatJ). In FAO Fisheries and Aquaculture Department [online]. Rome. Updated 2019. www.fao.org/fishery/ statistics/software/ fishstatj/en

FAO, Moffitt, C. M., & Cajas-Cano, L. (2014). Blue growth: the 2014 FAO state of world fisheries and aquaculture. *Fisheries (Bethesda), 39*(11), 552–553.

FDF (Federal Department of Fisheries). (2008–2015). *Fisheries statistics of Nigeria* (5th ed.). 52 pp.

Fernández, J. I. et al. (2011). Coastal Fisheries of Mexico. Coastal Fisheries of Latin America and the Caribbean. FAO Fisheries and Aquaculture Technical Paper. No. 544, FAO, pp. 231–284.

Fernando, D., Bown, R. M. K., Tanna, A., Gobiraj, R., Ralicki, H., Jockusch, E. L., ... Caira, J. N. (2019). New insights into the identities of the elasmobranch fauna of Sri Lanka. *Zootaxa, 4585*(2), 201–238.

Fernando, D., & Stewart, J. D. (2021). High bycatch rates of manta and devil rays in the "small-scale" artisanal fisheries of Sri Lanka. *PeerJ, 9*, e11994.

Ferretti, F., Worm, B., Britten, G. L., Heithaus, M. R., & Lotze, H. K. (2010). Patterns and ecosystem consequences of shark declines in the ocean. *Ecology Letters, 13*(8), 1055–1071.

Fricke, R. et al. (2021). Eschmeyer's catalog of fishes: Genera, species, references. California Academy of Science. Retrieved from http://researcharchive .calacademy.org/research/ichthyology/catalog/fishcatmain.asp.

Frisk, M. G., Miller, T. J., & Fogarty, M. J. (2001). Estimation and analysis of biological parameters in elasmobranch fishes: A comparative life history study. *Canadian Journal of Fisheries and Aquatic Sciences, 58*(5), 969–981.

Fromentin, J.-M., Bonhommeau, S., Arrizabalaga, H., & Kell, L. T. (2014). The spectre of uncertainty in management of exploited fish stocks: The illustrative case of Atlantic bluefin tuna. *Marine Policy, 47*, 8–14.

Gamba, Manoel da Rocha. (1994). *Guia Prático De Tecnologia De Pesca.* ICMBIO. Retrieved from https://www.icmbio.gov.br/cepsul/images/stories/ biblioteca/download/trabalhos_tecnicos/pub_194_gamba_guiapratico.pdf.

Hall, G. B., & Close, C. H. (2007). Local knowledge assessment for a small-scale fishery using geographic information systems. *Fisheries Research, 83*(1), 11–22. https://doi.org/10.1016/j.fishres.2006.08.015.

Hawkins, J. P., & Roberts, C. M. (2004). Effects of artisanal fishing on Caribbean coral reefs. *Conservation Biology, 18*(1), 215–226. https://doi.org/10.1111/j .1523-1739.2004.00328.x.

Heithaus, M. R., Frid, A., Wirsing, A. J., & Worm, B. (2008). Predicting ecological consequences of marine top predator declines. *Trends in Ecology and Evolution, 23*(4), 202–210.

Hind, D., Mountain, G., Gossage-Worrall, R., Walters, S. J., Duncan, R., Newbould, L., Rex, S., Jones, C., Bowling, A., Cattan, M., Cairns, A., Cooper, C., Goyder, E., & Edwards, R. T. (2014). Putting life in years (PLINY):

A randomised controlled trial and mixed-methods process evaluation of a telephone friendship intervention to improve mental well-being in independently living older people. *Public Health Research, 72*(2), 341–358.

Hobbs, J. E. (2019). Heterogeneous consumers and differentiated food markets: Implications for quality signaling in food supply chains. *Canadian Journal of Agricultural Economics/Revue Canadienne d'Agroeconomie, 67*(3), 237–249.

IBGE, Instituto Brasileiro de Geografia e Estatística. (2021). *Balneário Camboriú*. IBGE. Retrieved from https://cidades.ibge.gov.br/brasil/sc/balneario-camboriu/panorama.

ICMBio, Instituto Chico Mendes de Conservação da Biodiversidade. (2016). *PAN Tubarões: Plano de Ação Nacional para a Conservação dos Tubarões e Raias Marinhos Ameaçados de Extinção.* ICMBIO. Retrieved from https://www.icmbio.gov.br/portal/images/stories/docs-pan/pan-tubaroes/1-ciclo/pan-tubaroes-sumario.pdf

IUCN. (2021). *The IUCN red list of threatened species. Version 2021–3.* Retrieved February 16, 2022, from https://www.iucnredlist.org.

Jabado, R. W., Al Ghais, S. M., Hamza, W., Shivji, M. S., & Henderson, A. C. (2015). Shark diversity in the Arabian/Persian Gulf higher than previously thought: Insights based on species composition of shark landings in the United Arab Emirates. *Marine Biodiversity, 45*(4), 719–731.

Johnson, A. E., Cinner, J. E., Hardt, M. J., Jacquet, J., McClanahan, T. R., & Sanchirico, J. N. (2013). Trends, current understanding and future research priorities for artisanal coral reef fisheries research. *Fish and Fisheries, 14*(3), 281–292.

Karnad, D., Sutaria, D., & Jabado, R. W. (2020). Local drivers of declining shark fisheries in India. *Ambio – A Journal of the Human Environment, 49*(2), 616–627. https://doi.org/10.1007/s13280-019-01203-z.

Kerwath, S., Smith, C., Da Silva, C., Wilke, C. G. & Singh, L. (2013). *National plan of action for the conservation and management of sharks (NPOA-sharks).* Technical report: Department of Agriculture, Forestry and fisheries, Republic of South Africa, 68 pp.

Khazali, M. (2021). An overview of Persian Gulf environmental pollutions. *E3S Web of Conferences, 325*, 1–6.

Kilfoil, J. P., Campbell, M. D., Heithaus, M. R., & Zhang, Y. (2021). The influence of shark behavior and environmental conditions on baited remote underwater video survey results. *Ecological Modelling, 447*, 109507.

Krefft, G. (1968). Neue and erstmalig nachgewiesene Knorpelfische aus dem Archibenthal des Südwestantlantiks, einschließlich einer Diskussion einiger *Etmopterus*-Arten südlicher Meere. *Archiv für Fischereiwissenschaft, 19*(1), 42.

Krefft, G., & Stehmann, M. (1975). Ergebnisse der Forschungsreisen des FFS "Walter Herwig" nach Südamerika. XXXVI Zweitere neue Rochenarten aus dem Südwest-atlantic: *Raja (Dipturus) leptocauda* und *Raja (Dipturus) trachyderma* spec. nov. (Chondrichthyes, Batoidei, Rajidae). *Archiv. Fisherei Wissenchaft, 25*, 77–97.

Last, P. R., Seret, B., & Naylor, G. J. (2016). A new species of guitarfish, Rhinobatos borneensis sp. nov. with a redefinition of the family-level classification in the order Rhinopristiformes (Chondrichthyes: Batoidea). *Zootaxa, 4117*(4), 451–475.

Lessa, R. P., Santana, F. M., Rincón, G., Gadig, O. B. F., & El-Deir, A. C. A. (1999). *Biodiversidade de Elasmobrânquios do brasil. Relatório para o programa nacional de diversidade biológica (PRONABIO)–Necton–Elasmobrânquios, ministério do meio ambiente, dos recursos hídricos e da Amazônia legal (MMA)*. Recife: Ministério do Meio Ambiente, dos Recursos Hídricos e da Amazônia Legal (MMA).

López de la Lama, R., De la Puente, S., & Riveros, J. C. (2018). Attitudes and misconceptions towards sharks and shark meat consumption along the Peruvian Coast. *PLOS ONE, 13*(8), 1–16. https://doi.org/10.1371/journal.pone.0202971.

Marchetti, P., Mottola, A., Piredda, R., Ciccarese, G., & Di Pinto, A. (2020). Determining the authenticity of shark meat products by DNA sequencing. *Foods, 9*(9). https://doi.org/10.3390/foods9091194.

Martínez-Candelas, I. A., Pérez-Jiménez, J. C., Espinoza-Tenorio, A., McClenachan, L., & Méndez-Loeza, I. (2020). Use of historical data to assess changes in the vulnerability of sharks. *Fisheries Research, 226*, 105526.

Mascia, M. B. (2003). The human dimension of coral reef marine protected areas: Recent social science research and its policy implications. *Conservation Biology, 17*(2), 630–632. https://doi.org/10.1046/j.1523-1739.2003.01454.x.s

McEachran, J. D., & Compagno, L. J. V. (1980). Results of the research cruises of FRV "Walter Herwig" to South America. LVI: A new species of skate from the Southwestern Atlantic, *Gurgesiella dorsalifera* sp. November (Chondrichthyes, Rajoidei). *Arch. Fischereiwiss, 31*, 1–14.

MMA, Ministério do Meio Ambiente. (2006). *Avaliação do Potencial Sustentável dos Recursos Vivos na Zona Econômica Exclusiva do Brasil - Programa REVIZEE: Relatório Executivo.* Brasilia, Distrito Federal, Brazil: MMA.

MMA, Ministério do Meio Ambiente, (2021), *Biodiversidade. Ministério Do Meio Ambiente.* Retrieved from https://www.gov.br/mma/pt-br/assuntos/ biodiversidade.

Mojetta, A. R., Travaglini, A., Scacco, U., & Bottaro, M. (2018). Where sharks met humans: The Mediterranean Sea, history and myth of an ancient interaction between two dominant predators. *Regional Studies in Marine Science, 21*, 30–38. https://doi.org/10.1016/j.rsma.2017.10.001.

Mollet, H. F., & Cailliet, G. M. (2002). Comparative population demography of elasmobranchs using life history tables, Leslie matrices and stage-based matrix models. *Marine and Freshwater Research, 53*(2), 503–515.

Moore, J. E., Cox, T. M., Lewison, R. L., Read, A. J., Bjorkland, R., McDonald, S. L., … Kiszka, J. (2010). An interview-based approach to assess marine mammal and sea turtle captures in artisanal fisheries. *Biological Conservation, 143*(3), 795–805.

Musick, J. A., & Bonfil, R. (Eds.). (2005). *Management techniques for elasmobranch fisheries.* Rome, Italy: Food and Agriculture Organization of the United Nations.

Niedermüller, S. et al. (2021). *The shark and ray meat network: A deep dive into a global affair.* World Wide Fund for Nature. Retrieved July 21, 2021.

Nuñez, G. (2007). *Quality and stability of cuban shark liver oil: Comparison with icelandic cod liver oil.* UNU Fishers Training Program, Project, University of the United Nations, Fisheries Training Programme, p. 231

Pauly, D. (2006). Major trends in small-scale marine fisheries, with emphasis on developing countries, and some implications for the social sciences, 7–22.

Pauly, D., Christensen, V., Guénette, S., Pitcher, T. J., Sumaila, U. R., Walters, C. J., … Zeller, D. (2002). Towards sustainability in world fisheries. *Nature, 418*(6898), 689–695. https://doi.org/10.1038/nature01017.

Pauly, D., Watson, R., & Alder, J. (2005). Global trends in world fisheries: Impacts on marine ecosystems and food security. *Philosophical Transactions of the Royal Society B: Biological Sciences, 360*(1453), 5–12.

Pérez-Jiménez, J. C. et al. (2016). Las pesquerías artesanales de elasmobranquios como parte de sistemas pesqueros complejos en el sur del Golfo de México. *Ciencia Pesquera, 24*, 113–124.

Pérez-Jiménez, J. C., & Mendez-Loeza, I. (2015). The small-scale shark fisheries in the southern Gulf of Mexico: Understanding their heterogeneity to improve their management. *Fisheries Research, 172*, 96–104.

Prince, J. D. (2002). Gauntlet fisheries for elasmobranchs–the secret of sustainable shark fisheries. *Journal of Northwest Atlantic Fishery Science, 35*, 407–416.

Rincon, G. et al. (2017). Deep-water sharks, rays, and Chimaeras of Brazil. *Chondrichthyes - Multidisciplinary Approach*. https://doi.org/10.5772/intechopen.69471.

Rosa, R., & Gadig, O. B. (2014, November 25). Conhecimento Da Diversidade Dos Chondrichthyes Marinhos No Brasil: A Contribuição De José Lima De Figueiredo. *Arquivos De Zoologia, 45*, 89–104. https://doi.org/10.11606/issn.2176-7793.v45iespp89-104.

Rose, D. A. (1996). *An overview of world trade in sharks and other cartilaginous fishes*. TRAFFIC Network. Retrieved September 29, 2021.

Rubio-Cisneros, N. T., Moreno-Báez, M., Glover, J., Rissolo, D., Sáenz-Arroyo, A., Götz, C., … Herrera-Silveira, J. (2019). Poor fisheries data, many fishers, and increasing tourism development: Interdisciplinary views on past and current small-scale fisheries exploitation on Holbox Island. *Marine Policy, 100*(February), 8–20. https://doi.org/10.1016/j.marpol.2018.10.003.

Sadowsky, V. (1965). The hammerhead sharks of the littoral zone of São Paulo, Brazil, with the description of a new species. *Bulletin of Marine Science, 15*(1), 1–12.

Salas, S., Chuenpagdee, R., Seijo, J. C., & Charles, A. (2007). Challenges in the assessment and management of small-scale fisheries in Latin America and the Caribbean. *Fisheries Research, 87*(1), 5–16. https://doi.org/10.1016/j.fishres.2007.06.015.

Schlickmann, M. (2016). *Do Arraial De Bonsucesso a Balneário Camboriú: Mais De Cinquenta Anos De História* (1st ed.). Balneário Camboriú, Brazil: Fundação Cultural.

Seidu, I., Brobbey, L. K., Danquah, E., Oppong, S. K., van Beuningen, D., Seidu, M., & Dulvy, N. K. (2022). Fishing for survival: Importance of shark fisheries for the livelihoods of coastal communities in Western Ghana. *Fisheries Research, 246*, 106157.

Sethi, S. A., Branch, T. A., & Watson, R. (2010). Global fishery development patterns are driven by profit but not trophic level. *Proceedings of the National Academy of Sciences, 107*(27), 12163–12167.

Sheppard, C. (2016). Coral reefs in the Gulf are mostly dead now, but can we do anything about it? *Marine Pollution Bulletin, 105*(2), 593–598.

Shiffman, D. S., & Hammerschlag, N. (2014). An assessment of the scale, practices, and conservation implications of Florida's charter boat-based recreational shark fishery. *Fisheries, 39*(9), 395–407.

SL-NPOA-Sharks. (2013). Sri Lanka national plan of action for the conservation and management of sharks. Report by the ministry of fisheries and aquatic resources development, department of fisheries and aquatic resources, and the national aquatic resources research and development agency, 34 pp.

Sumaila, U. R., Zeller, D., Hood, L., Palomares, M. L. D., Li, Y., & Pauly, D. (2020). Illicit trade in marine fish catch and its effects on ecosystems and people worldwide. *Science Advances, 6*(9). https://doi.org/10.1126/sciadv.aaz3801.

Swain, D. P., & Benoît, H. P. (2015). Extreme increases in natural mortality prevent recovery of collapsed fish populations in a Northwest Atlantic ecosystem. *Marine Ecology Progress Series, 519*, 165–182.

Tanna, A., Fernando, D., Gobiraj, R., Pathirana, B. M., Thilakaratna, S., & Jabado, R. W. (2021). Where have all the sawfishes gone? Perspectives on declines of these critically endangered species in Sri Lanka. *Aquatic Conservation: Marine and Freshwater Ecosystems, 31*(8), 2149–2163.

Tesfamichael, D., Pitcher, T. J., & Pauly, D. (2014). Assessing changes in fisheries using fishers' knowledge to generate long time series of catch rates: A case study from the Red Sea. *Ecology and Society, 19*(1). https://doi.org/10.5751/es-06151-190118.

United Nations Office on Drugs and Crime. (2019). *Rotten fish: A guide on addressing corruption in the fisheries sector.* Retrieved September 29, 2021.

Vannuccini, S. (1999). *Shark utilization, marketing and trade.* No. 389. Food & Agriculture Org. Retrieved September 28, 2021.

Vooren, C. M., & Klippel, S. (Eds.). (2005). *Ações para a conservação de tubarões e raias no sul do Brasil,* 1st ed. Sandro Klippel, Porto Alegre, Brazil.

Wever, L., Glaser, M., Gorris, P., & Ferrol-Schulte, D. (2012). Decentralization and participation in integrated coastal management: Policy lessons from Brazil and Indonesia. *Ocean & Coastal Management, 66*, 63–72. https://doi.org/10.1016/j.ocecoaman.2012.05.001.

Worm, B., Davis, B., Kettemer, L., Ward-Paige, C. A., Chapman, D., Heithaus, M. R., … Gruber, S. H. (2013). Global catches, exploitation rates, and rebuilding options for sharks. *Marine Policy, 40*, 194–204.

Conclusion

Written by David Shiffman with contributions
by Catherine Macdonald, Lisa Whitenack,
Jasmin Graham and Camila Cáceres

DOI: 10.1201/9781003260370-5

Dr. David Shiffman

I've loved sharks ever since I was a kid—even though I grew up in Pittsburgh, Pennsylvania, pretty far from the ocean! I went to college at Duke University, majoring in Biology and Marine Science, and I got to spend time at Duke's marine lab in the Outer Banks and study abroad on the Great Barrier Reef at James Cook University. I went to the College of Charleston in South Carolina to get my MSc in Marine Biology, with a project studying the diet and ecology of sandbar sharks, and I got my PhD in Interdisciplinary Environmental Science and Policy at the University of Miami. As of this writing, I live and work in Washington, DC, where I am a Faculty Research Associate for Arizona State University. I study ocean conservation policy issues and teach introduction to marine biology at ASU. I'm also the author of *Why Sharks Matter* from Johns Hopkins University Press, and you can follow me on Twitter, Facebook, and Instagram @WhySharksMatter, where I'm always happy to answer any questions anyone has about sharks.

Dr. Catherine Macdonald

I'm an interdisciplinary environmental scientist who mostly studies both people and animals in tropical marine systems. I co-founded Field School (www.getintothefield.com), a marine field science training program with a focus on conducting safe, responsible marine field work in an inclusive and supportive environment. I have more than a decade of experience conducting field research on sharks and rays, and teaching animal handling and restraint. I'm also a Lecturer at the Rosenstiel School of Marine and Atmospheric Science at the University of Miami and a 2021 National Geographic Explorer. I'm proud to be a friend and volunteer mentor with MISS!

Dr. Lisa Whitenack

I grew up in the Chicago area, far from the ocean. I have a BSc in geology at the University of Illinois at Urbana-Champaign, where I discovered in my senior year that sharks left behind millions of teeth in the fossil record and many of them don't look like teeth from the sharks that live today. I kept studying fossil teeth during my MSc at Michigan State University, and then switched to studying living sharks for my PhD in Integrative Biology at the University of South Florida so that I could better understand how extinct sharks may have lived their lives. Today I am an Associate Professor of Biology and Geology at Allegheny College in Meadville, Pennsylvania. I'm far from the ocean again, but am within a few hours of three museums with great fossil sharks and shark jaw collections that allow my students and me to continue research on how shark teeth work.

TAKEAWAYS FOR ALLIES

There is a lack of diversity in science due to systemic obstacles which have excluded and continue to exclude BIPOC scientists and scientists from other marginalized communities. This in turn impacts the way scientists engage with community stakeholders and communicate science. When white (or cis, or male) classmates and colleagues first learn of these problems, many of them want to know what they can do to help. This section will introduce the concept of being an *ally*, someone who tries to use their privilege to lower barriers and make their workplace, community, or field a more welcoming, inclusive, and safe place for everyone. Describing someone as an "ally" isn't a title or identity—it's a term defined by action. Just because one has served as an ally in some situations or in the past doesn't mean they are an ally now, if they aren't acting like one. Remember that "being an ally" is a verb—an ongoing, active process—much more than "ally" is a noun. Allyship means following the lead of the people most impacted by the systemic barriers.

An effective ally uses their privilege to help their colleagues and co-workers. The word "privilege" sometimes provokes an emotional response in people who have it, as they may feel acknowledging that others face greater systemic barriers and challenges is saying they've "had it easy"—which of course is not necessarily true. Privilege doesn't mean someone has never faced any challenges or that they don't deserve their accomplishments, it just means that other people face additional barriers that they have not had to face for reasons that have nothing to do with individual merit. For example, consider a white, straight, cis male from an upper-middle class family, in their journey into shark science. They:

1. Had many available role models from similar backgrounds featured in media, the news, and your textbooks, and never had reason to doubt shark science was "for them"
2. Likely didn't face rejection or discouragement, or need to fear rejection or discouragement from peers or mentors because of their identity or identities
3. Experienced fewer physical risks of violence, or need to fear physical violence, as part of their schooling and training, including rape, sexual harassment, or hate crimes
4. Could afford to participate in the field opportunities that often help aspiring scientists begin their careers, either because they could afford to pay to participate or because they could afford not to work during the summers or breaks when these opportunities are most likely to be offered

Of course, none of this means this man's science journey was easy. It just means, on average, they faced fewer built-in barriers that could have prevented them from being able to continue along a certain path or succeed. Recognizing the intersection of various kinds of privilege can help provide perspective on the challenges people with less privilege might face. The more privilege a person has, the more responsibility they bear for trying to reduce systemic barriers for others.

To be an effective ally, a key first step is asking colleagues what they need instead of guessing or assuming, because those colleagues from marginalized communities know the problem much better than someone not impacted by the systemic issues (including what does not work in terms of solutions). That said, many scientists from underrepresented backgrounds are exhausted from having to answer the question "How can I help?" so many times in the past few years, so it's a good idea to do some reading and learning about the basics before you start engaging on specifics, as sometimes the answer to "How can I help?" has been stated repeatedly and publicly. If there is any doubt, donating to and raising money for organizations like Minorities in Shark Sciences (MISS) and Black in Marine Science (BIMS) will always be helpful. Some suggested further readings provide links from this book's website. It's very important to keep in mind that wanting to help is great, and trying to help is great, but neither is the same as actually helping—good intentions only go so far. It should never be assumed that someone's desire to help means that they are actually helping.

WHAT DO ACTS OF ALLYSHIP LOOK LIKE IN PRACTICE?

SPEAKING UP

Sometimes allyship means telling a mutual colleague that what they're doing or saying is problematic and/or harmful. That colleague may be more willing to listen to a person with perceived privilege, and that same message might be much less well-received if it came from someone from an underrepresented background. Sometimes, the colleague is not willing to listen at all. Remember to maintain open communication with the folks most impacted by the systemic issues. In some circumstances, it may be more harmful than helpful to call attention to a problematic statement or behavior in the moment. It is better to speak in support of the perspectives of those from marginalized communities and avoid speaking for them. To talk to a privileged colleague about their problematic behavior or statements, the conversation can often be framed around the discomfort and perception of the situation by not only the person from the marginalized community, but also by the person acting as an ally. Validating the experiences and perceptions of members of a marginalized community, both publicly and privately, can also be very helpful, as they may be told they're "too sensitive" or "taking it the wrong way" when they try to address problems. Doing things like speaking up can be stressful and exhausting, and may possibly have personal or professional repercussions. That work shouldn't fall exclusively to the people already most harmed by the problematic behavior.

MAKING SURE EVERYONE GETS THE SAME OPPORTUNITIES

Sometimes allyship means recommending a colleague or student for a job or other professional opportunity, and talking about how great they are in professional circles. This includes suggesting a colleague as someone who can be interviewed for a high-profile media opportunity or a public panel discussion, because representation matters—it makes a huge difference for kids to be able to see someone who looks like them doing a job. Folks should also be taking a look at the professional opportunities they're getting invited to and noticing who isn't there, and saying things like "hey it seems you have only invited white people to speak at this symposium, we should invite some colleagues of color." In the classroom, make sure to include marginalized voices in readings, in assignments, and in classroom discussions. And people shouldn't discount the value of regularly advocating for more diverse hiring practices at their institutions, though just bringing more people into an environment that's toxic for them doesn't do much to help if the organization doesn't also work at making the environment safer and more inclusive.

For folks in a position to invite guest speakers to their institutions, they can invite some of their colleagues from underrepresented backgrounds—noting here that they should be invited to talk about their research or conservation expertise, not just about their experiences with discrimination and diversity issues (unless they explicitly say that's what they want to talk about).

TAKING AN ANNOYING TASK OFF SOMEONE ELSE'S PLATE

Sometimes allyship means volunteering to help with a job around the workplace or in the classroom. This is especially important for people who identify as male, because lots of tedious tasks often fall to women and minorities in the office. Something as simple as taking a time-consuming, tiring, mildly annoying task off of someone else's plate can make a huge difference. Be the note-taker. Get the coffee. Offer to bring the snacks for a meeting. Offer to be a point person on scheduling. In addition to offering to take on these minor but time-consuming tasks, people can also recommend their BIPOC colleagues for the higher-profile or more professionally beneficial tasks.

LISTEN TO UNDERREPRESENTED VOICES (AND AMPLIFY THEM) ON SOCIAL MEDIA

For folks active on social media, following scientists from underrepresented backgrounds and sharing or re-tweeting their posts can help spread their messages (and followers often learn a lot themselves from these feeds). It's worth taking a few minutes for everyone to look at who they follow, and consider what voices they're not hearing. (This same dynamic can occur in department meetings or classrooms, as some colleagues or classmates may monopolize the conversation leaving little room for input from colleagues from underrepresented backgrounds, when they're even invited into the room at all). Additionally, please keep in mind that sometimes just being quiet can be an act of allyship. It is often important to sit and reflect on any discomfort a conversation has caused, because the (often defensive) first thought that pops into someone's head is unlikely to be an especially useful contribution to the discussion.

MENTOR AND OFFER ADVICE AND SKILLS TRAINING

People with skills that are professionally useful or people who have lots
of good professional advice should consider volunteering to mentor
colleagues from underrepresented backgrounds can also be a big help. In
academia, processes like graduate school applications and admissions are
often very unclear to people unfamiliar with them, the same is true of the
job market in environmental advocacy or policy spaces.

WHAT NOT TO DO

It's also important to note that there are some things that first-time
allies often try to do that do not actually help, or may even make things
worse. Some of these are done by people who are well-intentioned but
don't know any better, but they can still cause harm. Sometimes this
looks like highlighting someone's work with the caveat that "even though
they're <IDENTITY>," or adding someone to a committee or working
group as a token minority in a shallow effort to appear more diverse.
Minimizing someone's experience as minor or "not that bad" because it
seems unimportant, or giving advice when none is asked for, or trying
to solve problems without having a conversation with those affected are
all common mistakes people make. Centering oneself in a conversation
about diversity is another common mistake here. Also, while it's very
important to listen to colleagues from underrepresented backgrounds,
it is important to make sure that wounds are not being reopened by
making them relive a traumatic experience, and resist the (apparently for
some people) overpowering urge to correct details from someone's story
about how they were harmed.

While allies can help to dismantle or lower barriers, this should be
done to help colleagues from underrepresented backgrounds, not just to
make the supposed "ally" feel or look better. A common problem here
is white savior-ism, which is when a person does something that they
think will help, but in a way that implies that the people they're helping
would be lost without them, or a way that makes them the "hero" of the
story. There's also something called "performative allyship," which is

someone who pretends to help because they think it makes them look good. Performative allies don't actually want to do any of the hard work needed, especially if that work is not visible to others or they cannot expect to receive credit or reward for it.

CONCLUSION

Doing the work of allyship is difficult, time-consuming, stressful, and sometimes frustrating. It can be scary to speak up to those in power, and colleagues or classmates may cut ties with those standing up to the oppressive systems. There is no such thing as a perfect ally, and no one expects that—no matter how good people's intentions are, they'll mess up sometimes. It can be painful to admit failures, but they happen to everyone. The best anyone can do is apologize and do their best to make it right and learn from their mistakes. It is the commitment to learning and improving that makes an effective ally, someone who can contribute to helping to make the workplace, classroom, and the entire field better.

LEARN MORE

Whether you are a teenager looking for opportunities to join the professional field of marine biology, a seafood lover eager to learn more about science and conservation, a parent wanting to teach their children about the importance of sharks, an established professional looking to support innovative projects, or a volunteer willing to donate time, we have provided here a list of resources.

RESOURCES FOR YOUNG ADULTS

If you are a highschooler or college undergraduate looking to become a professional scientist, knowing the scholarships and grants to apply to will make a huge difference in your career. Oftentimes, minorities, immigrants, or first-generation students are unaware of sources available for financial and professional support to accomplish their goals. Here we have compiled a list of prestigious scholarships and grants that all newcomers should be aware of:

- Save Our Seas Foundation – Small Grant and Keystone Grants
  Save Our Seas is a non-profit organization based in Switzerland that funds shark and ray projects all around the world. They open their Small Grant (<$10,000 USD) and their Keystone Grant ($10,000–100,000 USD) every year during the spring. Make sure to check their website for more details and deadlines:
  https://saveourseas.com/grants/funding-applications/
  Important note: these grants are typically for MSc and PhD graduate students.
- National Science Foundation – Graduate Research Fellowship Program
  The NSF GRFP supports students in STEM disciplines with an annual stipend of $34,000 USD and an education allowance for research of $12,000 USD for three years. Receiving this fellowship is extremely competitive and prestigious, thus if you are awarded it you are pretty much guaranteed your choice of university and laboratory to join. This application is usually due in October, but make sure to check their website for deadlines.
  https://www.nsfgrfp.org/
  Important note: this fellowship is only for masters and doctoral students pursuing their education in a US institution. US citizenship or residency is required. Graduate students are only allowed to apply once, so make sure you have plenty of time and support to write and apply.
- Dr. Nancy Foster Scholarship Program
  Funded by NOAA, this program provides support for MSc and PhD degrees in oceanography, marine biology, or marine archaeology. Yearly support of up to US$42,000 per student (a 12-month stipend of US$30,000 in addition to an education allowance of up to US$12,000), and up to US$10,000 of support for a 4–6 week program collaboration at an NOAA facility. This

scholarship particularly encourages women and members of minority groups to apply. Application is usually due in December, but make sure to check their website for updated information.

https://fosterscholars.noaa.gov/howtoapply.html

Important note: US citizenship or residency is required. Make sure to find a collaborator at NOAA to write you a letter of recommendation, it will improve your chances of getting funded.

RESOURCES FOR SEAFOOD LOVERS, FISHERS, AND OTHERS IN THE MARINE INDUSTRY:

If you are a seafood lover, a fisher, a diver, a boat captain, involved in marine ecotourism, or any other industry that depends on marine or shark resources, sometimes it can be hard to get up-to-date information on scientific discoveries, policies, or news releases. Here, we have compiled a list of resources that are reliable and current.

- Seafood Watch

   Developed by the Monterey Aquarium in California, Seafood Watch is a sustainable seafood advisory list. It provides information on a regional scale (within the US and internationally) of seafood options labeled as "Best Choice," "Good Alternative," and "Avoid." Through their website and their app, it can help guide you to make responsible decisions when catching, purchasing, and eating seafood.

   https://www.seafoodwatch.org/recommendations
- Marine Stewardship Council (MSC)

   MSC is a global and international nonprofit organization that works to end overfishing around the world. After a long auditing and certification process, they provide "MSC stamps" to label sustainable seafood options that can easily be identified at the supermarket.

   https://www.msc.org/en-us/about-the-msc/what-is-the-msc
- US Fish and Wildlife Service

   US F&W is the federal agency in the United States in charge of conserving, protecting, and enhancing fish, wildlife, and their habitats for the benefit of the American people. Their main responsibility is the conservation and management of these resources, and thus all new policies, regulations, and laws regarding marine and shark conservation are overseen by them. To stay up to date on the latest on the conservation front, please visit their website. Please keep in mind that there may be additional policies and laws at the state level (in addition to the federal laws), so make sure to check your local Fish and Wildlife agency's website.

https://www.fws.gov/

RESOURCES FOR PARENTS AND THEIR CHILDREN

Teaching the next generation the importance of science, research, and conservation is of upmost importance to ensure the health of the oceans and the planet. Although many children may not live near the ocean or an aquarium, and may not learn about marine animals and conservation at school, there are many online resources available to learn about sharks.

- Sharks4Kids

    Sharks4Kids' goal is to create a new generation of shark advocates through a wide range of educational materials and experiences. On their website, you can find curriculums for teachers, coloring sheets, arts and crafts ideas, and much more.

    https://www.sharks4kids.com/
- Gill Guardians

    Content created to educate the general public about sharks, skates, and rays, the threats they face, and conservation efforts to protect them. Courses include video lessons, activities, quizzes, and action items. Our K-12 program will give students a chance to learn about shark biology and conservation while engaging with women of color working in the field of shark science.

    https://www.gillguardians.misselasmo.org
- Gills Club

    Gills Club is an education initiative dedicated to connecting young girls with female scientists from around the world. Anyone can become a Gills Club Member to learn more about sharks and how to become an ocean steward. From webinars, and a podcast, to occasional in-person events, learn more about how to join this club:

http://www.gillsclub.org/

SCIENCE COMMUNICATION RESOURCES

Whether you are looking to listen to a science podcast or books to read, we have compiled a list of our MISS' top picks.

- Podcasts
  1. Sharkpedia – Meghan Holst and Amani Webber-Schultz
  2. Conciencial Azul – Melissa Cristina Marquez
  3. The Whole Tooth – Dr Isla Hodgson

- Books
  1. *Shark Biology and Conservation: Essentials for Educators, Students, and Enthusiasts* – Daniel Abel and R. Dean Grubbs
  2. *The Shark Handbook: The Essential Guide for Understanding the Sharks of the World* – Greg Skomal
  3. *Shark Research: Emerging Technologies and Applications for the Field and Laboratory* – Jeffrey C Carrier, Michael R. Heithaus, Colin A. Simpfendorfer

MARINE CONSERVATION ORGANIZATIONS

Minorities In Shark Sciences (MISS) – www.misselasmo.org
Field School – www.getintothefield.com
Black in Marine Science (BIMS) – www.blackinmarinescience.org
Atlantic White Shark Conservancy – https://www.atlanticwhiteshark.org/
Bimini Biological Field Station – www.biminisharklab.com
American Shark Conservancy – https://www.americansharkconservancy.org/
Love the Oceans – https://lovetheoceans.org/
Oceans Research – www.oceans-research.com

Index